Reinforced concrete simply explained

JOHN FABER

B.Sc., C.Eng., F.I.C.E., F.I.Struct.E., M.Cons.E.
Consulting Engineer

and

DAVID ALSOP

B.Eng., C.Eng., M.I.C.E., M.I.Struct.E.

SIXTH EDITION

OXFORD UNIVERSITY PRESS

1976

Oxford University Press, Ely House, London W.1

OXFORD LONDON GLASGOW NEW YORK
TORONTO MELBOURNE WELLINGTON CAPE TOWN
IBADAN NAIROBI DAR ES SALAAM LUSAKA ADDIS ABABA
KUALA LUMPUR SINGAPORE JAKARTA HONG KONG TOKYO
DELHI BOMBAY CALCUTTA MADRAS KARACHI

ISBN 0 19 859514 X

Printed in Great Britain by
Fletcher & Son Ltd, Norwich

Preface

The forerunner of this book was written by Oscar Faber — my father — in 1922. At that time reinforced concrete was a relatively new material, shrouded in some mystery because of its composite nature. There had been no helpful literature for students or general readers, and *Reinforced concrete simply explained* (1922) was a notable success. In over fifty years it sold out five editions, the last of which I prepared 16 years ago, some time after my father had died.

Since that last edition there have been sweeping changes in concrete technology, including the appearance in 1972 of a very different Code of Practice, CP.110, *The structural use of concrete*. This switches from the long-established elastic theory of design, to an entirely fresh concept known as the limit state, in which the materials are considered in their inelastic condition just before reaching ultimate failure. Meanwhile we have changed in Britain from Imperial units to the SI units which are basically metric and similar to the European CGS system, though the unit of force has been taken as the newton — a force equal to the weight of a reasonable sized apple.

The new Code adopts a much more realistic attitude towards actual strengths of concretes; and the use of high-yield steels has become normal in most circumstances. The concept of the limit state has brought with it the introduction of partial safety factors applied simultaneously to materials and to loading conditions. The accepted approach to beam design — while still not strictly accurate — takes closer note of how the materials are likely to behave just before failure. The philosophy of shear resistance at last takes account of the longitudinal tensile strength of the member; and it relates particularly to the simpler details of construction generally adopted today. Column design has now become barely recognisable from how it was done 15 years ago using the old elastic theory.

Under all these changed circumstances it would have been unrealistic to try and produce a mere revision of the original book, and the present volume is offered as a complete re-write. In the task of preparing this I have had much help from my colleague David Alsop, who has the advantage of being closer to the everyday process of detail work carried out in the design office. In re-writing the book we have taken the

v

opportunity to say more about different forms of slab construction; to enlarge the chapters on prestressed concrete and water-retaining structures; to add a new chapter on precast concrete; and to include some detailed description of concrete itself in a final chapter on the anatomy of concrete.

However the book remains with the same title, and seeks as before to serve merely as an *explanation* of reinforced concrete. It is not a design manual; nor is it a slavish interpretation of any code, regulation or system. Despite the high degree of sophistication now current in modern reinforced-concrete practice, we have been at pains to keep the book simple and as near as possible to the easy style of its predecessor. It must of course be clearly understood that there is no pretence that the book should rank as highly technical, since clearly you cannot have it both ways; nevertheless I greatly hope that this modest volume may serve as a useful introduction or 'companion' for anyone launching out on a more detailed study.

It would have been easy to write a simple book that was full of half-truths or omitted reference to all complex considerations. We have tried our best to steer clear of such pitfalls. Our hope has been to show how simple designs can be made, safe but not necessarily the last word in economy. Where finer detail would have gone beyond the intentions of a simple book, we have at any rate given a few pointers to the sorts of direction an experienced designer would be likely to take.

Arthur Allen, M.A., B.Sc., F.I.C.E., of the Cement and Concrete Association, who has been a friend for over 25 years, has very kindly read the manuscript and made helpful comments for which I am extremely grateful. Warm thanks are also due to Alf Shain for help with preparing the diagrams, and to Miss Esther Harris for her patience and skill in preparing the typescript.

Harpenden, Herts   John Faber
September 1975.

Contents

1

Steel and concrete

Before we start discussing reinforced concrete, we need to understand
something about its two basic components: steel and concrete. We need
to know how these individual materials are comprised, and what are
their properties of strength and deformation under load; then we shall
be in a position to understand how they can be used functioning
together, each contributing qualities the other lacks in terms of
strength, economy, and other suitability.

Steels

Steel is not a pure metal. Ordinary mild steel contains about 99 per
cent iron, and of the remaining 1 per cent usually about half is manga-
nese, about a quarter is carbon, and the balance is silicon, sulphur, and
phosphorus. Of these minor constituents, carbon is the most important
from our point of view: the others originate as impurities, and provided
their content is kept down below certain limits they have no adverse
effect.

We come across three types of steel in ordinary reinforced concrete
work, as follows.
(1) *Mild steel bars* are produced directly from billets at a white heat
(1200°C) by rolling and re-rolling many times, all at great speed, until
we get down to the diameter of bar we require. Following the hot-
rolling process, the bars receive no further significant treatment.
(2) *Hot-rolled high-yield steel bars* are produced in similar fashion, but
are of considerably greater strength by virtue of skilful adjustment of
the chemical composition of the steel before the formation of the
billets. This steel may have a higher carbon content than ordinary mild
steel, though not necessarily so.
(3) *Cold-worked high-yield steel bars* are produced by taking mild steel
bars (as (1) above), and subjecting these at ordinary temperatures to
considerable physical deformation, which results in the steel strain-
hardening and developing thereby a much increased strength.

The measurement of strength of different steels is difficult to

1

express in simple terms, because other qualities of suitability arise; nevertheless we shall come to recognize what is meant by the steels (1), (2), and (3) above having respectively strengths of 250, 410, and 425 N/mm^2.

To understand the behaviour of *mild steel* under load, let us follow now in detail the procedure of testing a mild steel bar in the laboratory. We take a bar, say 25 mm diameter, and grip the two ends in special jaws which are so arranged that the more we pull on the bar the tighter the jaws grip. The jaws form part of a very powerful testing machine which enables us to pull the bar in tension by drawing the jaws further and further apart until eventually the bar will break. This increasing of the load on the bar is done fairly slowly, and we shall watch carefully all the time to see what is happening to the bar.

Two marks are made on the bar, a distance apart equal to five times the bar diameter: thus with our 25 mm bar, the marks are 125 mm apart. We now start loading the bar, and up to a stress† of 250 N/mm^2 or more there is no stretching of the bar visible to the naked eye; but then suddenly the bar elongates appreciably, though still continuing to carry competently the same load. This is the part B C of curve (1) in Fig. 1.1. This curious, sudden and very obvious stretching of the steel at constant load is known as yielding, and the stress at which it occurs is known as the yield stress.

In the process of yielding, another surprising property of steel comes into play, namely that the considerable deformation of stretching has the effect of hardening the steel and making it stronger than it was before we first started loading it. This phenomenon is known as cold-working. After yielding a certain amount (perhaps about 5 per cent), sufficient strain-hardening has occurred to enable the steel to go on and carry more load than previously. Thus we proceed up the part of the curve between C and D, with the bar continuing to stretch but all the time increasing in strength at a greater rate than the accompanying reduction of its cross-sectional area. At the point D, the point of maximum load, we find the effect of drawing the jaws of the testing machine further apart is to cause somewhere a local reduction in the

---

†*Stress* means load per unit area, so with our 25 mm bar, which has a cross-sectional area of 491 mm^2, the force in the bar corresponding to 250 N/mm^2 would be 250 x 491 = 123 000 N, i.e. 123 kN. This is roughly equal to the force which gravity exerts on a mass of 12·3 tonnes. In common usage the two are frequently interchanged, and without harm; although strictly speaking the reader should realise that kilonewtons are a *force* and tonnes are a *mass*.

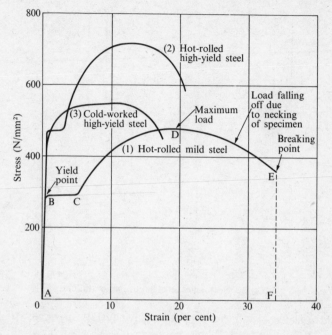

FIG. 1.1. Stress–strain curves for various steels, stressed in tension at constant rates of strain.

diameter of the bar (known as necking, see Fig. 1.2) with the result that the bar can now only carry less load; the intensity of stress at the position of the neck becomes excessive, and suddenly the bar breaks at this position with a sharp report, as represented by the point E on the curve.

We said that up to 250 N/mm^2 or more there was no visible elongation of the bar. This is true. However if very delicate measuring apparatus is attached to the bar, it can be detected that the bar is actually stretching over this range, though only by a very small amount; and provided we do not load the bar up to yield stress, subsequent removal of the load will allow the bar to recover immediately to its original length. This is what is known as elastic behaviour, and up to the yield point the bar is said to be stressed within its elastic range. This is the range within which we limit the stress in our steel reinforcements for design purposes. But it is a comfort and important to note that beyond the yield point the bar still has a considerable margin of safety within its maximum load-carrying capacity; and before breaking it

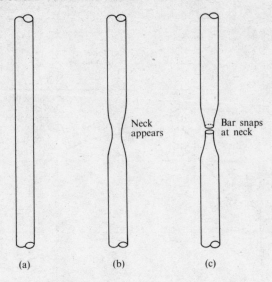

FIG. 1.2. Necking of steel bar at tension-failure.

would in fact stretch by an amount roughly equal to 35 per cent of its original length. So much for our test on mild steel.

If we now do a second test, this time on a bar of *hot-rolled high-yield steel*, we find that the strength and straining characteristics are similar in form to those of the mild steel, but different in degree. (See curve (2) in Fig. 1.1). Note that the yield point is now 410 N/mm^2 or greater, an increase of about 65 per cent; but the breaking strain† is only about two-thirds that for our mild steel bar, though for reinforced concrete design purposes still handsomely adequate.

Let us now give attention to the *cold-worked high-yield steel* represented by curve (3) in Fig. 1.1. This is the steel produced by strain-hardening ordinary mild steel by the technique of cold-working, i.e. by exploiting the phenomenon that we saw between points B and D on curve (1) when testing our mild steel bar.

It is clear from curve (3) that the breaking strain of the cold-worked steel is only about half that of the mild steel, and not surprisingly so since the mild steel had already been strained considerably in being cold-worked. The other point we notice from curve (3) is that there is

† *Strain* of our bar in tension means elongation per unit length. Thus strain = elongation/original length.

no marked yield point, and this again is not surprising, since with hot-rolled steels it is the yielding which achieves the cold-worked hardening, and having already once gone through the process of strain-hardening the same physical adjustment cannot occur a second time.

In the absence of a definite yield point for cold-worked steel, it is not so obvious how we should arrive at the limit of the range within which this class of steel should be used for design purposes. The accepted stress, comparable to the yield point, is known as the 0.2 per cent proof stress, which means the stress which will cause a 0.2 per cent permanent strain. For cold-worked steel bars over 16 mm diameter this value is 425 N/mm^2, and for bars 16 mm and under the value of 460 N/mm^2 is normal. However we shall see later that cold-worked steel is not much used for smaller bar diameters, the exception being bars used in slabs. For simplicity throughout this book we shall stick to the one value of 425 N/mm^2 for cold-worked steel, any inaccuracy then always being on the side of safety.

The reader may well ask how can it be that curve (3) for the cold-worked steel shows a higher maximum load than for the ordinary mild steel curve even at D, bearing in mind that the cold-worked steel comes directly from the strain-hardening of the mild steel? In simple terms there are two main causes for the increase. The first is that in cold-working the amount of strength increase depends on the degree of disturbance of the crystalline structure of the steel; and in our simple test specimen the strain-hardening was merely unidirectional, whereas in the commercial production of cold-worked steel the strain-hardening is achieved by multidimensional deformations which are mainly torsional, but include also a certain amount of tensile strain as well. The second cause is that our tensile test was carried out over a matter of minutes only, whereas commercial cold-worked steel is allowed a period of many days over which to age-harden, through which time the structure of the steel is able to settle down to a more stable condition of enhanced strength. Further, of course, we have to remember the stress values at curve (1) were based on the diameter of the original mild steel bar before the onset of the cold-working deformations.

Having now shown the behaviours of the three types of steel, and their different strengths respectively of 250, 410, and 425 N/mm^2, we need to consider which of these steels we shall prefer to use in our reinforced concrete work. Up till the time of the Second World War, ordinary mild steel bars were in general use for most reinforced concrete work; then through the fifties, steel manufacturers were able to reduce the relative cost of hot-rolled high-yield steel so as to make its greater strength attractive in relation to its price, and the tendency increased to move away from ordinary mild steel to hot-rolled high-yield steel. However the hot-rolled high-yield steel involved the manufacturers in producing two steels of different chemical compositions; and more recently there has been a tendency to limit the manufacture of steel to ordinary mild steel quality, and subsequently convert this to high-yield strength by cold-working. Thus today it has become general to use cold-worked steel at 425 N/mm^2 for all main longitudinal reinforcing bars; but for links, which are of smaller diameter and have to be bent easily to relatively small radii, it is normal to use the soft mild steel with its yield strength of 250 N/mm^2. These are the two grades of steel we shall be referring to throughout this book.

These strengths of 250 N/mm^2 and 425 N/mm^2 are what we term the *characteristic strengths* of these steels, and we denote this by the symbol f_y. Strictly speaking the characteristic strength of a steel is the yield point or 0.2 per cent proof stress below which not more than 5 per cent of all the bars tested will fail. And as a further safeguard it is required that none of the bars tested would ever yield or have a 0.2 per cent proof stress less than 0.93 of the specified characteristic strength.

Real structures are actually designed to a stress less than the characteristic strength of the steel so as to provide an insurance against unforseeable circumstances. This is discussed further in Chapter 2 when we talk about partial safety-factors for strength.

Concretes

Concrete is a mixture of small stones, sand, and cement, which, when mixed with water, and left to stand, sets and forms a hard mass like rock. Like natural rocks, concrete is both hard and brittle. Its strength depends much on the proportions of the mixture, and also on its age since the time water first came into contact with the cement. The notable strength of concrete is in compression; its strength in tension is very much less, something of the order of one-tenth.

The increase in strength of concrete is most marked over the first few weeks, and slows down thereafter. At 28 days, concrete has already reached three-quarters of the maximum strength it will attain even if left for many years; and this is convenient because it helps make rapid progress in construction, enabling us to rely on the strength of the concrete without further waiting. It is the 28-day strength of the concrete we take into account in all our design calculations.

Generally speaking, the more cement there is in the mixture, the stronger the concrete will be; however cement is expensive, and for ordinary purposes it is found unnecessary to make the concrete stronger than a given amount. The quantity of water is also important: too much water reduces the strength of the hardened concrete, and too little water prevents the concrete being compacted into a dense mass. The correct quantity of water to be used depends on the type of stone being used but is usually a little more than half the weight of the cement.

For convenience of small-scale working, cement can be bought in standard 50 kg bags, and the bagfull can then be used as a unit of measurement; so we arrive at a simple mix of suitable strength as follows:

cement	50 kg
sand	85 kg
stone	160 kg
water	28 kg.

For larger works, cement is delivered in bulk, and it is normal then to think in terms of material quantities that will produce a cubic metre of finished concrete. Thus the same mix as given above can be expressed as:

cement	360 kg
sand	600 kg
stone	1150 kg
water	200 kg.

Mixes arrived at in this simple way are known as *prescribed mixes*. They represent a fairly rough and ready approach to the proportioning of the materials, and are not as sparing of cement as mixes which have been designed more carefully and take into consideration the actual qualities of the sand and stone, and other details. This is discussed more later.

If some of our mixture, as defined above, is formed very carefully into cubes 150 mm across, and left 28 days to harden, the average crushing strength of the cubes would be found to be of the order of 33 N/mm^2. Certainly none of the cubes would be likely to be weaker than 25 N/mm^2. Indeed if any were as low as 21 N/mm^2, something would be wrong with the cement or the sand or the stone, or the mixing arrangements, or the way we were making the cubes or testing them; and we should have to stop and review the situation. Generally speaking, these 150 mm cubes of our prescribed mix are not likely to have a 28-day strength less than 25 N/mm^2, and for this reason the mixture we have defined is known as Concrete Grade 25.

25 N/mm^2 is what we term the *characteristic strength* of this concrete mix, and we denote this by the symbol f_{cu}.

It is possible, and indeed more usual, to produce Grade 25 concrete with a smaller proportion of cement than that indicated above for our prescribed mix, but then there needs to be some history of experience of the materials being used, as well as the mixing machinery and the supervision of its use. Such mixes are then known as *designed mixes*. Not only do designed mixes have the advantage of a smaller quantity of cement; the closer control of the materials and machinery enables us to accept the concrete without demanding such a high average strength as the 33 N/mm^2 referred to above for a prescribed mix, because we can anticipate less scatter of test results. If there has been six months' evidence of the materials and machinery and supervision, we need demand only an average crushing strength of about 29 N/mm^2; and if there has been twelve months' evidence, we need demand only an average crushing strength of 27 N/mm^2. Here we are talking of fairly sophisticated mixing procedures, as undertaken by the large readymix companies whose vehicles these days are such a familiar sight on the roads.

But whether we use a prescribed mix or a designed mix, we can ensure by testing our cubes that the 28-day strength of the concrete we shall have in our structure is the desired 25 N/mm^2; that is to say it really will be Concrete Grade 25. This is the strength of concrete we shall be giving our attention to throughout this book.

Other concrete grades behave in exactly the same sort of way, but the margins are then of course somewhat different. The other grades in normal use are 20, 30, and 40. Grade 20 generally leads to rather bulky beam and column sizes in superstructures; and in many soil conditions is too readily susceptible to attack from even modest sulphate concentrations. Concrete Grade 40 tends to be used only in rather special circumstances, where for example it is desirable to avoid bulky column sizes at the lower storey-heights of a fairly tall building. Hence the Grade 25 we have chosen comes within the centre range of concrete in most frequent use today; and for the expenditure of a reasonable quantity of cement should enable consistently good cube results to be achieved without undue difficulty.

The making up of 150 mm cubes is normal routine procedure on all real construction jobs in Great Britain and many other countries in Europe; and this is generally done about daily on each site. After hardening, the cubes are taken out of their moulds and then sent away to a laboratory and crushed at 28 days' age; and this enables us to check that the materials and mixing procedures on the job do not fall below the standards we have relied upon when preparing our designs.

Cubes are a convenient compact shape for casting, and storing and handling. They are made in machined cast-iron moulds of very precise dimensions. At the laboratory they are crushed between the two horizontal steel plates of a testing machine, the cubes now being arranged on their sides, so that the surface which was the top when the cube was being cast is now vertical. The purpose of this is that the top surface of the cube at casting can only be finished off with a trowel, and this inevitably is not as true as the other faces of the cube which are formed against the sides of the mould.

It is known by actual experiment that when a brittle material like concrete is crushed in a testing machine, the stocky shape of the cube gives consistently higher strength results than the same concrete would display in the more slender circumstances of a real beam or column. This is because in practical use the failure of concrete in compression is related to induced lateral tensile strains; and the loading plates of the testing machine tend to restrain such lateral expansion of the concrete, and the effect of this is particularly marked in units as stocky as cubes. More slender prisms are free to fail naturally without the same restraining effects from the loading plates. (See Fig. 1.3).

For this reason, other countries, notably the United States, Canada, New Zealand, and France make their standard test specimens as cylinders 150 mm diameter by 300 mm long. A further advantage claimed of the cylinder is that the crushing stress at its mid-height is more likely to be uniform across the horizontal plane. The difficulty with cylinders is to get a true surface on the top when the cylinder is being cast; and this is avoided with cubes which we have noted are always tested on their sides. Cylinders also suffer the disadvantage of greater bulk and storage inconvenience.

When cubes and cylinders are cast of identical concrete, it is known that cylinders generally fail at only about three-quarters the crushing load of the cubes. Concrete in real beams or columns is believed to fail

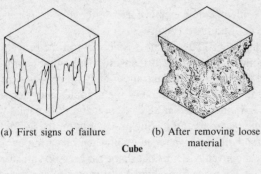

(a) First signs of failure   (b) After removing loose material

Cube

Cylinder

FIG. 1.3. Normal modes of failure of concrete cubes and cylinders.

at about two-thirds the cube strength. This does not matter, so long as we are clear in our minds that when we refer to a concrete having a characteristic strength (f_{cu}) of 25 N/mm^2, this relates to its performance in a cube, and its behaviour in a real structure will always be different and its strength only about two-thirds. This is discussed further in Chapter 2 when we talk about partial safety-factors for strength.

Now let us consider the stress/strain relationship of our concrete in just the same way as we did earlier for steel. A typical stress/strain curve for a Grade 25 concrete is shown in Fig. 1.4. Note here that the scales of this curve are altogether different from the scales of our stress/strain curves for steel as given in Fig. 1.1. Whereas steel is just starting its yield process when the strain has reached about 0·3 per cent, most concretes are so 'unyielding' and hard that when they have been strained about 0·3 per cent, they are reaching the point of incipient brittle failure.

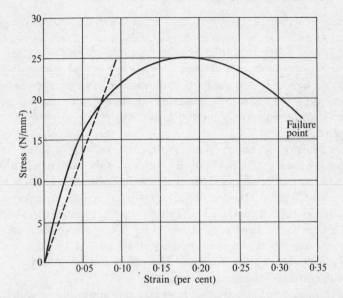

FIG. 1.4. Grade 25 concrete in compression at constant rate of strain (short term).

The stress/strain curve given at Fig. 1.4 is typical for short-term loading over a period of five minutes or so, as would occur in a normal compression strength test. In Chapter 3 we shall see that when concrete is loaded persistently over a period of several months or years, the strains increase to about three times the short-term values due to what is known as creep; but at present we need not concern ourselves with this. What is important to note about the shape of the concrete stress/strain curve is that there is no clearly defined elastic range over which strain is proportional to stress; nor is there anything which compares at all with the yielding we saw in the test on our steel bar. This is typical of any brittle material, and is one of the great short-comings of concrete. It is of course the reason why concrete needs to be reinforced with steel for most practical uses, other than where its compressive strength can be taken advantage of on its own, as in certain massive foundations, gravity retaining walls, dams, and similar structures.

Perhaps in completing this chapter we could now note some comparisons between concrete and steel, as follows. The *ultimate* compressive strength of our 25 N/mm² concrete is only about 6 per cent of the *0·2 per cent proof stress* of our 425 N/mm² steel. Furthermore concrete fails abruptly and without warning at a strain of about 0·3 per cent; whereas steel on the other hand gracefully accommodates strains many times greater than this, giving plenty of warning before approaching the onset of distress, still continuing to carry its load competently with a margin of strength in hand. It should also be noted that whereas steel itself is roughly equally strong in both tension and compression, concrete in tension is only about one-tenth as strong as it is in compression.

As a matter of interest, referring to Fig. 1.1, the area ABCDEF under the curve (1) for mild steel is a measure of the work that has to be done on a unit area of the material in order to break it. In the cases of curves (2) and (3) the comparable areas and hence the work done are clearly less but still considerable. If the reader will compare these with the area under the curve for concrete in Fig. 1.4, noting of course the extreme differences of scales between Figs. 1.1 and 1.4, he will appreciate graphically the great difference there is between the strength characteristics of steels and concrete.

Having said this, however, we must take care not to end up with the impression that concrete does not suffer strain at all. This would be wrong, as Fig. 1.4 so clearly shows; and in Chapter 4 we shall see that these very small strains of concrete within a reinforced concrete member allow the concrete to make inelastic adjustments of stress, and this has an important and useful influence on the ultimate strength of the member.

2

Partial safety factors

In Chapter 1 we explained why we decided to restrict our attention throughout this book to Grade 25 concrete, and normally to 425 steel. We can take it that Grade 25 concrete will not fail in compression in a 150 mm cube at any stress below 25 N/mm^2; and we know the 425 steel will be within its elastic range up to a stress of 425 N/mm^2.

Suppose now we place a real load of 650 kN on a short concrete column 200 mm x 200 mm; then the compressive stress in the column will be

$$\frac{650 \times 10^3}{200 \times 200} = 16 \cdot 2 \text{ N/mm}^2,$$

which relates to a stress in a 150 mm cube of

$$\frac{16 \cdot 2}{0.67} = \underline{24 \cdot 5 \text{ N/mm}^2}.$$

And if we pull a 25 mm steel bar with a load of 200 kN, the tensile stress in the bar will be

$$\frac{200 \times 10^3}{491} = \underline{400 \text{ N/mm}^2}.$$

Both these stresses are less than the characteristic strengths of the respective materials, so the concrete should not fail in compression and the steel should still be within its elastic range.

However in practice we would never seek to stress our materials so close to their characteristic strengths, because this would allow no margin for errors or other unforeseen circumstances. Such circumstances can arise from the actual loads turning out to be greater than we had thought, or the materials being weaker than we had suspected, or a combination of both. Unfortunately we know from long experience that however much care is taken in design and construction under

13

practical conditions, unforeseen contingencies keep creeping in, and the only way we can limit the number of failures of real structures to the relatively low proportion that actually occur is to apply factors of safety that give us margins of security over (a) the loads we make provision for, and (b) the materials we rely on to carry those loads. Let us review a few such contingencies.

Consider for example an ordinary office building. A suitable thickness for the floor slab might be 150 mm, and on this basis it is not difficult to calculate what the floor would weigh; however if mistakes of construction occur so that the slab is actually built 15 mm thicker than intended and therefore 10 per cent heavier, it is important there should be a margin in the beams supporting the slab so that these can cope satisfactorily with the overload involved. Similarly, there may be an allowance in the design of the floor for it to carry partitions made of materials having a certain weight; and if at a later date the number of partitions is increased or the materials used for them are heavier than had been allowed for, it is important there should be sufficient margin in the strength of the floor to prevent it collapsing as a result.

Often it is easier to assess the loads of the building itself (known as the dead loads) than it is to predict the loads imposed subsequently when the building is put into service. For example a reasonable intensity of imposed loading for which to design a normal office floor would be $2 \cdot 5$ kN/m^2. However ten persons occupying a room 6 m x 6 m would represent a loading of only $0 \cdot 25$ kN/m^2, and general furniture would probably be only about the same amount again; yet on the other hand if six hefty men were squeezed together onto an area 1 m square (as can happen in a crowded situation) they would represent a loading of 6 kN/m^2. From one such simple example it is clear there is no unique answer for the imposed load for which we should design a floor, knowing for sure that this will cater for all eventualities and still be within the bounds of reasonable economy. Suppose it were desirable later to put a heavy safe in the room: should we have been wise if we had designed the floor without allowance to make this possible? And what if in installing the safe it were accidentally dropped off the trolley whilst wheeling it in? Of course with proper care and attention we should be able to avoid the trolley mishap; and probably, with a little thought, we can arrange to share the load of the safe over an area of floor which is not likely to be loaded much otherwise: but clearly if floors were to collapse each time they were the least bit overloaded or abused, there would be many more structural disasters than we know to be the case today. It is for reasons such as these that factors of safety have to be applied to all our best estimates of loads.

Now let us switch our attention to the materials. The Grade 25 concrete we tested in our 150 mm cube was probably treated with a fair amount of respect in terms of the thoroughness with which the wet materials were compacted into the mould, and no doubt a good deal of care was taken to see the cube was stored free from physical disturbance, or extremes of drying wind and sunshine, or freezing temperatures. Can we be so sure that on a busy construction site the concrete will receive the same care in every vital part of our structure? The chain will only be as strong as its weakest link. Can we be sure the sand and stone will always be as clean as in our test specimen, and the cement as fresh; and can we be sure the proportions of the ingredients may not get out of adjustment at some time, or wood shavings or a cigarette packet fall into the formwork and become part of the finished work?

Similarly with the steel. Occasionally a batch of bars slips through the system untested, or failure to reach the characteristic strength goes by unnoticed — or unheeded. What if a few of the bars in rolling at the mills finish with slightly less than the cross-section intended? What if the bars have rusted by being stored a year or two outdoors? These are all reasons for applying a factor of safety to the characteristic strengths of our materials.

Other errors tend to creep in to most real jobs, however much engineers seek to prevent it. These can be errors of design calculation, or mistakes in preparing the drawings or reinforcement schedules. Sometimes these arise from careless slips; sometimes from misunderstanding. How many columns, for example, have been designed as though they were centrally loaded when in fact allowance should have been made for considerable eccentricity and bending?

Then there are general errors of workmanship. Bars in a slab may be required to be spaced at 150 mm intervals; yet perhaps the steel-fixer gets muddled by looking at a wrong drawing and as a result spaces the bars at 200 mm intervals. This would have the effect that the stress in the bars would be 33 per cent greater than we had intended when we were preparing the design.

And through all the contingencies mentioned above there is our overall ignorance or uncertainty of how reinforced concrete really works within an individual member, how the concrete and the steel actually share out the loads and stresses between themselves. In

Chapter 4 we try to argue rationally what is going on, but whether the concrete and steel are sympathetic to all our theories is another matter! All one can claim is that the calculations engineers make today are based on bold and simplifying assumptions which lead to results experience has shown are normally satisfactory.

One of the simplifications we make is to limit our number of factors of safety to two, one related to the *loads*, and one related to our *materials strengths*. As these two factors of safety are always involved together, they are both known as *partial safety-factors*.

Partial safety-factor for loads

Let us deal first with the partial safety-factor for loads. This is denoted by the symbol γ_f, so we have

$$\text{design load} = \gamma_f F_k, \tag{2.1}$$

where F_k is the *characteristic load*†. This means that after we have made our best possible estimate of the characteristic load we *multiply* it by the partial safety-factor (γ_f), so that the load we actually design for is *greater* than the characteristic load, giving us an additional margin of security. Different partial safety-factors apply in different circumstances.

In the present book we shall be restricting our attention to

characteristic dead loads (G_k), and
characteristic imposed loads (Q_k).

The respective design loads are then taken as

$$\text{design dead load} = 1 \cdot 4 \, G_k, \tag{2.2}$$

$$\text{design imposed load} = 1 \cdot 6 \, Q_k, \tag{2.3}$$

where $1 \cdot 4$ and $1 \cdot 6$ are the appropriate partial safety-factors (γ_f). Note that the imposed-load factor $1 \cdot 6$ is greater than the dead-load factor: this is because of the greater uncertainty that arises in determining suitable values for imposed loads, as we discussed earlier in our example of the floor in an office building.

In the examples we shall be giving throughout this book, we shall assume that the dead loads (which can include floor slabs, walls, partitions, and all finishes) are equal to or greater than the imposed loads. This then enables us, for convenience, to take a compromise

† *Characteristic load* means the best realistic estimate we can make of the load, believing it unlikely that this would ever be exceeded significantly.

value of 1·5 for our partial safety-factor for loads, and any error arising from such simplification will be on the side of safety. In practice, when doing an important design, one would of course apply separately the two factors 1·4 and 1·6 to the dead and imposed loads respectively, or at any rate check that the two loads were approximately equal.

Partial safety-factor for materials

The partial safety-factor for materials is denoted by the symbol γ_m. For safety's sake we need our design calculations to be on the assumption that our material strengths might be *less* than the characteristic strengths, so we use our factor γ_m by *dividing* it into the characteristic strengths (f_k) of the materials.
Thus we have

$$\text{design strength} = \frac{f_k}{\gamma_m}. \tag{2.4}$$

For concrete, bearing in mind the sudden abrupt way it fails, and all the practical problems that arise over mixture proportions, compaction, weather, dirt, and similar, the value taken for γ_m is 1·5. And remembering from Chapter 1 that the concrete in a real structure is only about two-thirds of the 28-day cube strength (i.e. $0·67 f_{cu}$), we have

$$\text{concrete compressive design strength} = \frac{0·67 f_{cu}}{1·5}$$

$$= 0·45 f_{cu}. \tag{2.5}$$

This means that with our Grade 25 concrete we shall work to a compressive design strength of $0·45 \times 25$ N/mm^2, i.e. 11·2 N/mm^2.

For steel, which is manufactured under very much more closely controlled conditions, and which has the reassuring property of ample stretching long before approaching failure, it is appropriate to apply a lesser value for γ_m to the tensile yield or 0·2 per cent proof stress. This value is taken as 1·15. Thus we have

$$\text{steel tensile design strength} = \frac{f_y}{1·15}$$

$$= 0·87 f_y. \tag{2.6}$$

With our 425 steel this means we shall be working to a tensile design strength of $0·87 \times 425$ N/mm^2, i.e. 370 N/mm^2.

Combined effect of the safety-factors

It is interesting now to look back and see where these equations have brought us.

The load safety-factors of 1·4 and 1·6 at equations (2.2) and (2.3) often combine to average out at about 1·5: and the minimum materials safety-factor was 1·15 for the steel reinforcement as given at equation (2.6). Thus we have a least margin overall to protect us against errors of one type or another of 1·5 x 1·15 = 1·73. This means we have an insurance against distress which ranges between (a) a margin of 73 per cent for errors in assessing our loads assuming everything else in the construction is correct, or (b) a margin of 43 per cent for errors in our materials assuming the assessment of design loads has been correct. This is reassuring.

Actually the margins are somewhat greater than this, as we shall see in Chapter 4: certain inelastic adjustments will take place between the concrete and steel in the limit state before physical collapse of a practical structure could ever occur.

Comment

Equation (2.5) for concrete in compression is correct when precise calculations are done. However with certain short-cut design techniques, values less than 0·45 have to be used, and these are explained in the chapters that follow. In calculations of shear strength (which are really to do with inclined tensions) very much reduced stresses have to be worked to, as discussed in Chapter 6.

Equation (2.6) for steel is correct for bars used in tension. However for bars used in compression, the design strength is governed by considerations of buckling. For our 425 steel we have

$$\text{steel compressive design strength} = 0.74\,f_y, \qquad (2.7)$$

giving in our case a compressive design strength of 0.74×425 N/mm^2, i.e. 315 N/mm^2. We shall see later that short-cut devices affect the design of compression steel in columns much as mentioned above for concrete.

The significance of the differences from the factors derived at equations (2.5) and (2.6) will all slot into place when we work our way through a number of practical examples in the remainder of the book.

3

Reinforced concrete

In Chapter 1 we saw the very different properties of steel and concrete. Steel is strong in both tension and compression, and can accommodate very large strains. Concrete is nothing like so strong in compression, and in tension is weaker still; furthermore concrete is 'unyielding', which means the same thing as being brittle. Steel therefore has advantages over concrete.

Concrete on the other hand has many advantages over steel, the foremost of these being that the material itself is very much cheaper. It also has the great merit of permanence, saving the need for cost of painting or any other surface covering. Nor does concrete require special protection against risk of destruction by fire, whereas steel in most circumstances is unacceptable unless suitably clad by sprayed or pre-molded coverings, or indeed by casing in lightly reinforced concrete. A further great attraction of concrete over steel is that it is cast simply in moulds to any desired shapes and dimensions, enabling adventurous architectural and structural requirements to be met in the most direct and effective manner, whereas frequently with structural steelwork the extra costs of special fabrication, delivery, and erection to meet such demands would be exorbitant.

The success of reinforced concrete has been achieved by exploiting the best of various virtues of these two component materials. Certain fortunate compatibilities have helped make this possible, as for example their similar coefficients of expansion under changes of temperature, the good grip and adhesion that occurs when the concrete hardens around the reinforcements, and other matters of detail we shall be discussing later on. As a start, let us consider first the case of a simple beam.

Fig. 3.1 shows an unreinforced concrete beam of rectangular cross-section slung between two supports and loaded at the centre. The beam bends and takes up a curved shape; and realizing that the ends remain square to the axis, it is clear from the fact that the top has shortened and the bottom lengthened that the upper part of the beam

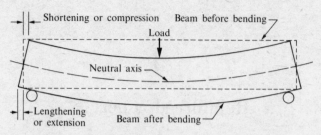

FIG. 3.1. Unreinforced concrete beam in simple bending.

must be in compression and the lower part in tension. Somewhere roughly midway between the top of the beam and the bottom there is a neutral axis which is neither compressed nor extended, and has remained its original length, and is therefore unstressed.

If our beam is made of concrete which will actually fail in compression at about 20 N/mm²†, we can assume that its tensile strength will be in the region of only about 2 N/mm². Accordingly the beam will fail when the tension stress at the bottom reaches 2 N/mm². The compression stress at the top will then also be only about 2 N/mm², so that the greater compressive strength of the concrete has not been brought into play, and the beam is therefore heavier than necessary and wasteful.

We can make the beam much stronger by strengthening the bottom, by inserting in it steel bars which will supply the tensile strength in which the concrete is deficient, until the tension side is now strong enough to bring into action the full strength of the compression in the top. Roughly the beam will then be ten times as strong as before. Up to this limit, the concrete is not fully stressed; beyond it additional steel will not increase the carrying capacity of the beam proportionately because the stress in the concrete will then have become the limiting consideration.

So far so good! But now we have to go back and review what is happening to the bottom part of the beam, the part under the neutral axis, as indicated in Fig. 3.1. We have just said that the concrete here cannot stand more than a certain amount of tension because it will fail. What is there to stop the concrete from cracking while it is still trying to take the tension, and before it gives up and passes the responsibility

† This would relate to a cube strength of 30 N/mm², which is the sort of real result we might get in practice using a Grade 25 concrete intended to have a minimum strength of 25 N/mm².

on to the steel? The answer is that nothing can prevent the concrete below the neutral axis cracking, and this in fact is what happens. The remainder of this chapter will be devoted to demonstrating this and understanding the significance of it.

Going back to Chapter 1, we remember from Fig. 1.1 that when we were testing our steel bar within the elastic range it elongated a very small amount; and the amount of this elongation was directly proportional to the stress in the bar. This is Hooke's Law, which says quite simply that within the elastic range *stress is proportional to strain.* We can express this by the formula

$$E = \frac{\text{stress}}{\text{strain}}, \tag{3.1}$$

or

$$E = \frac{\frac{P}{A}}{\frac{l}{L}} = \frac{PL}{Al}, \tag{3.2}$$

where E = a constant depending on the material (known as the elastic modulus or Young's modulus),
P = the load in the bar (N),
A = the cross-sectional area of the bar (mm^2),
L = the original length of the bar (mm),
l = the elongation of the bar (mm).

Whatever dimensions we take for A and L the relationship exists that E for steel always works out to have the same constant value of 200 kN/mm^2, and this is true for both mild steel and high-yield steel.

Thus if we have a bar of 425 steel stressed to (say) 400 N/mm^2, we know from equation (3.1) the bar will stretch so that

$$\text{strain} = \frac{\text{stress}}{E},$$

$$= \frac{400 \text{ N/mm}^2}{200 \times 10^3 \text{ N/mm}^2},$$

$$= 0.002.$$

If the original length of the bar was one metre, the elongation will be

$$0.002 \times 10^3 = 2 \text{ mm}.$$

Alternatively, our knowledge of E enables us to say, from a measurement of elongation, how great the stress must be which produces it. For example, if we are told a steel bar one metre long is elongated 2 mm by a tensile stress, we can say at once from equation (3.1) that the stress must be such that

$$\text{stress} = E \times \text{strain}$$
$$= (200 \times 10^3) \text{ N/mm}^2 \times 0 \cdot 002$$
$$= 400 \text{ N/mm}^2.$$

And if the cross-sectional area of the bar is (say) 800 mm^2, which is approximately the area of a 32 mm dia bar, the tension force on the bar must have been

$$400 \text{ N/mm}^2 \times 800 \text{ mm}^2 = 320\,000 \text{ N}$$
$$= 320 \text{ kN}.$$

Now a similar relationship exists with regard to concrete. If a specimen of concrete is subjected to tension, it also elongates. The elongation is again extremely small, but measureable with delicate apparatus: and although we saw in Chapter 1 that concrete does not behave in quite the same convenient linear fashion as steel, it is clearly possible with our Grade 25 concrete to visualize a compromise linear relationship as shown dotted on Fig. 1.4. Thus we have again the law of equation (3.1), that stress is proportional to strain, but for concrete the elastic modulus E has a different value which can be regarded in simple terms as 13 kN/mm^2.

Actually the determination of an E value for concrete is relatively vague and no fixed value exists. With short-term loading (5 minutes or so) the E value can be seen from Fig. 1.4 to be about $(20/0 \cdot 00075)$ N/mm^2 = 26 kN/mm^2. But if the stress persists, the strain will increase over a period of a year or so, until eventually it becomes roughly three times the amount measured on initial short-term loading. This increase of strain over a period of time is what is known technically as creep. With the incidence of creep, E falls to a value of about 9 kN/mm^2, or even less. Thus we see the figure of 13 kN/mm^2 is a good convenient middle-of-the-road value which is commonly adopted when considering average concrete strains. It will be noted that this is only one-fifteenth the E value of steel.

In other words, the elongation of concrete for a given stress is about fifteen times as great as for steel; or, conversely, if two bars, one concrete and one steel, are stressed so that they elongate the same amount, the stress in the steel will be fifteen times as great as that in the concrete. This is an extremely important result, which helps us visualize the relationships between stress and strain in a reinforced concrete member, so long as the materials are operating within their elastic ranges. We can express it another way as follows. The ratio between the elastic modulus of steel and the elastic modulus of concrete has the value of 15; so that

$$\frac{E_s}{E_c} = 15. \tag{3.3}$$

Now when a beam of concrete having steel bars in it is stretched, one of two things must happen. Either the steel elongates exactly the same amount as the concrete immediately surrounding it, or the rods slip in the concrete: and it will have to be accepted at this stage that in good design such slipping is adequately guarded against by the feature of *bond* which we shall be discussing later. Hence, in general, it may be taken that the steel elongates the same amount as the concrete surrounding it: and from equation (3.3) we see the stress in the steel will be fifteen times as great as that in the concrete surrounding it.

It follows therefore that if the steel is stressed to (say) 370 N/mm^2 in tension (see equation (2.6)), the concrete stress would be

$$\frac{370}{15} = 25 \text{ N/mm}^2 \text{ in tension.}$$

But, as previously explained, our concrete breaks in tension when it is stressed to only 2 N/mm^2. Therefore with the steel stressed to 370 N/mm^2, the concrete will definitely have cracked in tension, and is therefore incapable of resisting tensile stresses. *Hence the steel must be designed to carry the whole tension, receiving no assistance from the concrete.*

Do not suppose that the cracks referred to above do not exist because they cannot be seen with the naked eye. They are generally so fine that they cannot be detected, as is clear from the following. The elongation of the steel bars stressed to 370 N/mm^2 is such that

$$\text{strain} = \frac{370}{200 \times 10^3}$$

$$= 0.0018,$$

so that if cracks are (say) 100 mm apart, their width will be only 100 x 0.0018 = 0.18 mm, and if closer, proportionately less.

In practice it is considered that crack-widths up to about 0·3 mm are acceptable and unlikely to affect the durability or appearance of the structure in normal circumstances. However the variability of E_c (largely due to creep) and the uncertainty of the distance apart of the cracks obviously makes precise calculation of crack-widths impossible, particularly so when the complication of shrinkage and temperature effects are taken into account. Generally a designer guards against excessive cracking by meeting rule-of-thumb minimum requirements for reinforcement percentages and spacings. These will be referred to later.

Looking back now to Fig. 3.1 (exaggerated as it may be), it is clear some sort of watch needs to be kept that our beams, however strong, do not deflect too much. Deflections not exceeding 1/250th of the span are generally acceptable. With bad design it is possible to devise a beam too shallow (or a slab too thin) so that these will *deflect excessively*, yet by virtue of a very great tension force provided by the reinforcement be *adequately strong*. This can come about either by the steel being too high a percentage of the beam's cross-section, or by the stress in the steel being to a high design strength ignoring its *stretch* in relation to the depth of the member.

Deflection amounts are much influenced by the span of the member, and by the amount of freedom there is for it to slope at its end supports. In modest building structures, with beams and slabs of the sorts of duties and dimensions given in the examples in this book, deflection worries are unlikely to arise. Any risk is most likely to occur in adventurous designs where cold-worked high-yield steel is worked up to its maximum design stress in thin slabs carrying heavy loads on long spans.

The Code of Practice CP.110 (1972) *The Structural Use of Concrete†* gives full guidance over the matters of crack control and deflection. Detailed recommendations are made there for bar spacings to prevent cracks; and tabulations are given for depths for beams and slabs to prevent excessive deflections, depending on the spans, support conditions, and percentages of steel reinforcement. If the reader feels he wants to follow these matters further, he is referred to the Code itself.

†CP.110 (1972) is the current British Standard Code of Practice covering the structural use of concrete. With its publication have been introduced many fresh concepts in terms of concrete design and practice from the time of the three earlier Codes CP.114, CP.115, and CP.116, which it has superseded. Copies of CP.110 can be obtained from British Standards House, 2 Park Street, London W1A 2BS at the published price.

Up to this stage of the book, we have indicated the basic principles underlying the design of reinforced concrete. Now we must go on to study in some detail how engineers believe the stresses in beams and slabs and columns are distributed, and from this how we can calculate the sizes required for such members, and what quantities of steel reinforcement will be needed to meet strength requirements. This is what the following chapters are about.

4

Rectangular beams

In Chapter 3 we considered the bending of an unreinforced concrete beam, and saw that the beam failed long before the full compressive strength of the concrete had been brought into play. We found the beam would be very much stronger if we introduced steel bars in the bottom to carry the tension force there. Here was our introduction to reinforced concrete.

Beam loaded in stages

Now let us do a laboratory study of the simple rectangular reinforced concrete beam, 500 mm x 300 mm, shown in Fig. 4.1, spanning 6·0 metres between supports and carrying a load uniformly spread along its length. The significant reinforcements are the three 25 mm dia bars in the bottom.† The concrete is Grade 25 (f_{cu} = 25 N/mm^2) and the steel is 425 cold-worked (f_y = 425 N/mm^2). The size and span of the beam are typical of what might be used in any modern reinforced-concrete building.

As soon as it is placed on its supports, we notice (using proper

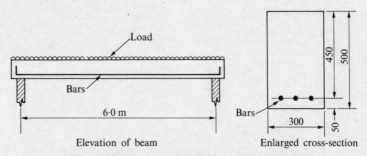

FIG. 4.1. Simple rectangular reinforced-concrete beam.

† In practice such a beam would also have vertical links spaced along its length to resist shear (see Fig. 6.9). There would also be two bars in the top to hold the links in place while the beam was being cast. But for the purpose of our present study we are only interested in the effect of the three bars in the bottom.

measuring apparatus) that the beam bends slightly, even under the action of its own weight. This is indicated as Stage A in Fig. 4.2, where, for clarity, the deflection of 1 mm is drawn exaggerated about 100 times. The associated stress diagram represents the stresses in the beam materials at midspan: the neutral axis occurring where the stress is zero;

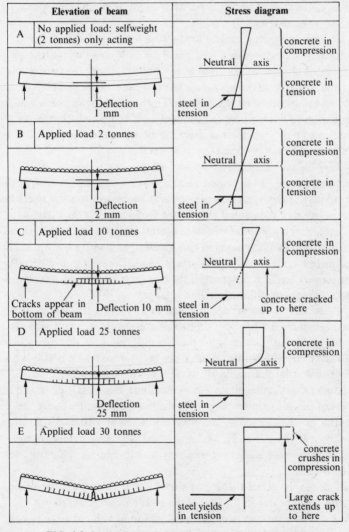

FIG. 4.2. Rectangular reinforced-concrete beam in bending.

the material above the neutral axis (having shortened) being in compression, and material below the neutral axis (having lengthened) being in tension. At this stage the tensile stress in the concrete is so small that the concrete has not cracked, and the stress diagram is linear (both materials operating within their elastic ranges) so that the maximum stresses are at the extreme top and bottom of the beam. The stress in the steel is 15 times the stress in the adjoining concrete as noted by equation (3.3).

If now we apply a 2 tonne load to the beam, we see that the deflection increases (Stage B of Fig. 4.2). Note also that a slight but important change has occurred in the stress diagram: the concrete has approached the maximum *tensile* stress it can sustain, and no longer behaves elastically near the bottom. This is the stage at which a similar unreinforced beam would fail suddenly and completely; but in our reinforced beam the onset of cracking merely causes the tension to be transferred from the concrete to the steel.

Now we increase the applied load in stages to 10 tonnes. The deflection is seen to increase to about 10 mm, and the cracking in the lower part of the beam extends upwards by degrees to somewhere near the position of the neutral axis (Stage C of Fig. 4.2). Note that this minute cracking is easier to detect nearest midspan where the effects of bending (the bending moment) are greatest, and where the stresses in the materials are consequently greatest too. The steel is now carrying all the tension, and the concrete all the compression; both materials are functioning well within their elastic ranges.

When we increase the applied load to 25 tonnes, the deflection of the beam increases to about 25 mm. The concrete is now stressed to about 17 N/mm^2, which is the maximum it can be relied upon to sustain (0.67×25 N/mm^2); and from Fig. 1.4 we realise that it is no longer behaving elastically, although it has the capability to accommodate further strain before finally failing. This it does; and the greatest possible compressive force the beam can muster above the neutral axis is achieved by the concrete taking up a stress distribution as shown at Stage D of Fig. 4.2 and at the same time extending the compression zone slightly downwards, i.e. the neutral axis moves down somewhat.

This has the effect that the line of the concrete compression force comes down nearer to the line of the steel tension force, so the lever-arm between the two becomes less and of reduced effectiveness. The concrete no longer has any reserves of strength or strain to draw upon, and all depends on what margins remain in the steel. At about 30 tonnes the steel reaches the end of its elastic range, and at midspan stretches visibly causing a crack to open up several millimetres wide at the bottom of the beam, extending well upwards and approaching the top. The stress distribution changes dramatically (Stage E, Fig. 4.2) with the compressive force now being taken on a rapidly reducing depth of concrete, until the stress in the concrete exceeds its crushing strength, and the concrete shatters. At this stage (unless there is some means of preventing the load pursuing the beam) the deflection increases suddenly until the beam slips off its supports and collapse is total.

Concrete stress block

It is important to have understood the behaviour of a reinforced concrete beam through the various stages of loading, as described above, in order to follow the modern concept of design technique known as *limit state*. Under this method an efficient beam is regarded as one which reaches a condition somewhere between Stages C and D when it is carrying its design load.[†] The steel will clearly be behaving elastically, not having reached its 425 N/mm^2 0·2 per cent proof stress; but we can envisage the concrete as straining disproportionately from its stress in accordance with the shape of the curve given at Fig. 1.4. Thus we have a concrete stress distribution as indicated at Fig. 4.3(a), where the maximum value of stress at the top of the beam is 0·45 f_{cu} as given by equation (2.5). For our Grade 25 concrete, this means a maximum concrete design stress of 11·2 N/mm^2.

Whilst such a stress diagram may have some academic appeal, it is sometimes convenient (particularly when doing calculations without the use of design charts) to think of this as being replaced by the simpler equivalent stress distribution shown in Fig. 4.3(b), taking a constant value of stress of 0·40 f_{cu}. Obviously this is easier to summate when arriving at the total compressive force; and it is easier too to determine then where such compression force acts. The configuration given in Fig. 4.3(b) is known as a *rectangular stress block*, and generally gives almost the same numerical answer as we get from using Fig. 4.3(a), saving much time and trouble and risk of error when design charts are not available.

[†] We have already seen that the *design load* is about 1·5 times the characteristic load, so we have a margin here. There are of course also margins provided in the materials stresses we shall be working to in our design calculations.

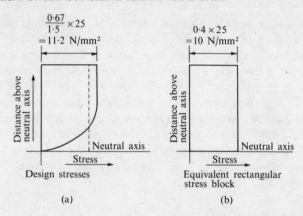

FIG. 4.3. Distribution of stress in the compression zone.

Whichever stress distribution the designer chooses to adopt, it is important that he limit the depth of the compression zone to a distance down from the top of the beam equal to half the depth to the tension steel. This is to avoid the risk of getting into the situation we saw happen as Stage D developed into Stage E.

Basic design theory

To follow the principles of design theory, we shall for simplicity now consider the stress diagram in Fig. 4.4 for a singly reinforced beam at the ultimate limit state, adopting the rectangular stress-block approximation. The concrete design stress is limited to $0.4 f_{cu}$, which for Grade 25 concrete is $0.4 \times 25 = 10$ N/mm². The total compressive force is denoted by C, and its line of action is through the centre of the rectangular stress block; the total tensile force is denoted by T.

These two opposing forces are of equal magnitude to one another (since otherwise there would be a resultant force tending to move the whole beam in the direction of its length) and their distance apart is z; consequently they set up a moment in the beam of $C \times z$ or $T \times z$. This is exactly the same principle as occurs in the action of a lever, where the effectiveness depends on the magnitude of a force multiplied by the perpendicular arm at which it acts. The distance z is known as the *lever arm*; and the moment within the beam is known as the beam's *resistance moment*.

There is a maximum resistance moment any beam can summon before failure, and this clearly depends on the values of C and T and z.

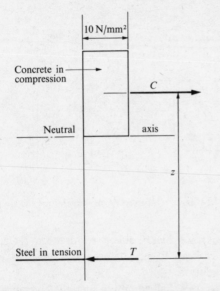

FIG. 4.4. Simplified stress-diagram for the whole cross-section.

But this maximum resistance moment only develops in the beam when it is carrying its ultimate load just prior to collapse. When the beam is only partly loaded, the resistance moment will be less than the maximum, so that C and T will be correspondingly less also.

To design a beam we need somehow to relate the effect of the load on it to the resistance moment; and in Chapter 5 (see Fig. 5.1) we see that when a load bends a beam it produces in the beam a *bending moment* which, like a resistance moment, is made up of a force multiplied by a perpendicular distance. In the case of a bending moment the force is the load (or loads) on the beam, and the distance depends on how long the beam is and where the load or loads come.

If failure of the beam is to be avoided, the resistance moment must always balance exactly the bending moment applied. If the beam is strong enough to do this, the beam is satisfactory; if it is not, the beam will collapse. In the laboratory study of the beam we did earlier, the resistance moment at Stage D (25 tonnes) was balancing the applied bending moment satisfactorily; at Stage E (30 tonnes) the bending moment was greater than the maximum resistance moment the beam could develop, and this is why the beam collapsed.

We shall now develop, in algebraic terms, the ultimate resistance moment of a singly reinforced beam such as we did our test on. Then

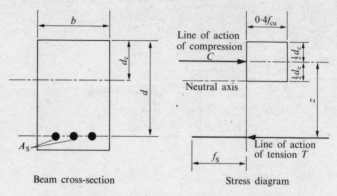

FIG. 4.5. Singly reinforced beam: dimensions and stresses.

for design purposes we shall be able to relate this to the applied bending moment in any real case to see whether the beam we have chosen is adequate for its duty.

For our purpose we shall adopt the simplified rectangular stress block as we approach the ultimate condition; and we shall use the notation indicated in Fig. 4.5, where

b is the chosen beam breadth;
d is the chosen effective depth of the tension steel,
A_s is the chosen area of tension steel,
f_s is the stress in the steel,
T is the total tension force,
d_c is the depth of concrete in compression,
C is the total compression force,
z is the lever arm.

Now the total tension force in the steel at any time is arrived at simply by multiplying the steel stress by the area of the steel, so that

$$T = f_s \times A_s. \tag{4.1}$$

This force enables the beam to resist a bending moment M such that

$$M = f_s \times A_s \times z. \tag{4.2}$$

Similarly the total compression force in the concrete is

$$C = (0 \cdot 4 f_{cu}) \times (b \times d_c), \tag{4.3}$$

which enables the beam to resist a bending moment M such that

$$M = (0.4\,f_{cu}) \times (b \times d_c) \times z. \tag{4.4}$$

If the bending moment M applied to the beam is increased in stages, the forces T and C will increase correspondingly. The only way T can increase (see equation (4.1)) is by an increase of f_s, since A_s is fixed. And at the ultimate limit the only way C can increase (see equation (4.3)) is by an increase of d_c, since f_{cu} and b are already determined.

Therefore for our beam to remain within safe design limits, M can only go on increasing until either

(i)   the steel reaches its design strength of $0.87\,f_y$ (equation (2.6))

or (ii)   the compression zone extends down to half the effective depth, i.e. so that $d_c = \tfrac{1}{2}d$.

When either of these conditions is reached, our applied bending moment M will have reached the ultimate resistance moment M_u of the beam. Thus M_u is the lower of two values, one depending on the steel, and the other depending on the concrete.

From consideration of the steel,

$$M_u = T \times z = (0.87\,f_y) \times A_s \times z. \tag{4.5}$$

And from consideration of the concrete,

$$M_u = C \times z = (0.4\,f_{cu}) \times (b \times d_c) \times z,$$

where d_c is now $\tfrac{1}{2}d$,
and z is $d - \tfrac{1}{2}d_c = d - \tfrac{1}{4}d = \tfrac{3}{4}d$,
so that

$$M_u = (0.4\,f_{cu}) \times (b \times \tfrac{1}{2}\,d) \times \tfrac{3}{4}\,d$$
$$= 0.15\,f_{cu} \times bd^2. \tag{4.6}$$

If f_{cu} is not to exceed 25 N/mm^2, we see from equation (4.6) that M/bd^2 is not to exceed 0.15×25 N/mm^2, i.e. not to exceed 3.75 N/mm^2. $\tag{4.7}$

It will interest the reader to know that equations (4.5) and (4.6) are the equations for the ultimate resistance moment of singly reinforced beams by the rectangular stress block technique, as given in the Code of Practice CP.110 (1972).

We are now equipped to assess the suitability of any singly reinforced beam to fulfil whatever bending duty may be required of it. There are two alternative approaches we can adopt. One is to pursue

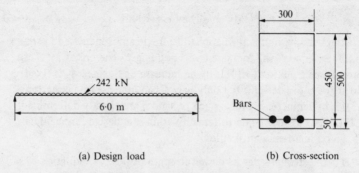

(a) Design load   (b) Cross-section

FIG. 4.6. Single-span rectangular reinforced-concrete beam.

the rectangular stress block technique, using equations (4.1) to (4.6) inclusive we have just derived; the other is to take advantage of the standard design chart given in Fig. 4.7, which is based on the more sophisticated concrete stress distribution given in Fig. 4.3(a). We shall work through an example both ways, and see how the answers we get compare, and what are the relative amounts of labour involved.

Example

Let us design by calculation the reinforced concrete beam we studied at the start of this chapter. See Fig. 4.6. The beam is to carry a total applied load of 14 tonnes spread uniformly over a simple span of 6·0 metres; the 14 tonnes is made up of 8 tonnes dead load and 6 tonnes imposed load. Our experience leads us to suppose a suitable beam size would be 500 mm deep by 300 mm wide; but our calculation will have to check that the concrete, at this size, will not be overstressed; and then we shall design an appropriate area of steel reinforcement for casting in the bottom.

The characteristic loads are:

self weight of beam: 0·5 m x 0·3 m x 6·0 m x 24 kN/m³ = 21·6 kN

applied load: 14 tonnes   =   approx   140·0 kN

                                              161·6 kN

Applying our safety-factor for loads, we have

design load $U = 161 \cdot 6 \times 1 \cdot 5$ $= 242$ kN

From Fig. 5.1(d) we have

applied bending moment $\quad M = \dfrac{UL}{8}$

$$= \frac{242 \text{ kN} \times 6 \cdot 0 \text{ m}}{8}$$

$$= 182 \text{ kN m.}$$

Method 1: Rectangular stress block method working from our equations

The reader will find this method exceedingly laborious, but it will complete his understanding of the theory of design. Method 2, which follows, is recommended for normal use.

First we can check quickly that the concrete section is big enough, so as to be sure our f_{cu} of 25 N/mm^2 is not exceeded. This is done from equation (4.7).
We have

$$\frac{M}{bd^2} = \frac{182 \times 10^3 \times 10^3 \text{ N mm}}{300 \times 450^2 \text{ mm}^3}$$

$$= 3 \cdot 01 \text{ N/mm}^2.$$

This is less than $3 \cdot 75$ N/mm^2, so we know our beam size is adequate.

Since M/bd^2 is considerably less than $3 \cdot 75$ N/mm^2, the neutral axis of our beam will not be at the half-way depth; so our next task is to locate it, i.e. to calculate d_c. Using equation (4.4), and knowing that

$$M = 182 \times 10^3 \times 10^3 \text{ N mm,}$$

$$f_{cu} = 25 \text{ N/mm}^2,$$

$$b = 300 \text{ mm,}$$

$$z = d - \frac{d_c}{2} = 450 - \frac{d_c}{2},$$

we have $\quad M = (0 \cdot 4 \, f_{cu}) \times (b \times d_c) \times z$

$$182 \times 10^3 \times 10^3 = (0 \cdot 4 \times 25) \times (300 \times d_c) \times \left(450 - \frac{d_c}{2}\right).$$

This reduces to

$$122 \times 10^3 = d_c (900 - d_c)$$

or
$$0 = d_c{}^2 - 900 d_c + 122\,000.$$

This awkward quadratic equation has now to be solved either by the binomial theorem, or by a process of trial and error: Either way is a considerable labour, but the answer arrived at is

$$d_c = 166 \text{ mm},$$

which, being less than half d, confirms our early quick check that the beam size is adequate and our characteristic concrete stress f_{cu} of 25 N/mm^2 will not be exceeded.

It remains now to calculate the area of steel required. Using equation (4.5), we know that

$$M = 182 \times 10^3 \times 10^3 \text{ N mm},$$

$$f_y = 425 \text{ N/mm}^2,$$

$$z = d - \frac{d_c}{2} = 450 - \frac{166}{2} = 366 \text{ mm}.$$

Thus

$$M = (0\cdot87 f_y) \times A_s \times z$$

$$182 \times 10^3 \times 10^3 = (0\cdot87 \times 425) \times A_s \times 366$$

$$\text{giving } A_s = \frac{182 \times 10^3 \times 10^3}{0\cdot87 \times 425 \times 366}$$

$$= 1344 \text{ mm}^2.$$

By referring now to the Table 4.1 of steel areas, we find this is conveniently provided by 3 bars of 25 mm diameter (area = 1470 mm^2).

Method 2: Using standard Design Chart (Fig. 4.7)

Fig. 4.7 is based on the more elegant stress distribution given at Fig. 4.3(a), and is suitable for use with any singly reinforced beam of Grade 25 concrete and 425 steel. The chart is taken from Part 2 of CP.110 (1972), where other curves will also be found, suitable for use with different concrete and steel stresses.

TABLE 4.1
Areas of steel bars (mm²)

Bar dia (mm)	Number of bars									
	1	2	3	4	5	6	7	8	9	10
6	28·3	56·6	84·9	113	142	170	198	226	255	283
8	50·3	101	151	201	252	302	352	402	453	503
10	78·5	157	236	314	393	471	550	628	707	785
12	113	226	339	452	566	679	792	905	1020	1130
16	201	402	603	804	1010	1210	1410	1610	1810	2010
20	314	628	943	1260	1570	1890	2200	2510	2830	3140
25	491	982	1470	1960	2450	2950	3440	3930	4420	4910
32	804	1610	2410	3220	4020	4830	5630	6430	7240	8040
40	1260	2510	3770	5030	6280	7540	8800	10100	11300	12600

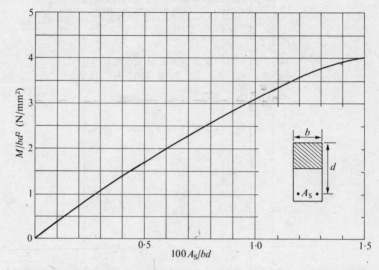

FIG. 4.7. Standard design-chart for a singly-reinforced concrete beam. (Grade 25 concrete and 425 steel).

By this method all we have to do is calculate M/bd^2 as before, equal to 3·01 N/mm²; then find the position 3·01 on the M/bd^2 scale, and read across to strike the curve, which we find is at the point where, on the other scale,

$$\frac{100\,A_s}{bd} = 0.98,$$

from which

$$A_s = \frac{0.98 \times b \times d}{100}$$

$$= \frac{0.98 \times 300 \times 450}{100}$$

$$= 1320 \text{ mm}^2.$$

If in reading across we find that our M/bd^2 value is too high to strike the curve at all, then we know the concrete section we have chosen is inadequate for our purpose.

A comparison of the two methods shows immediately which is the easier and less likely to lead to arithmetical slips. By using the Design

Chart we are always slightly closer to the true stress/strain characteristics of real concrete, though the difference between the answers obtained in our example is seen to be negligible. The purpose of once working our way through, in Method 1, using the equations we have derived earlier was merely to show to a conclusion the basic fundamentals of the theory we had been developing. In practice a designer would almost certainly adopt the Design Chart method, unless perhaps he preferred to work from tables which he could prepare for himself from the equations once and for all.

Indeed for any practical problem there is always an infinite number of beam sizes from which a choice can be made, and each would require a different area of steel reinforcement. Sometimes the beam dimensions are dictated by architectural considerations, but frequently they are determined by a balance of costs as between concrete, steel, and formwork, with an eye also on achieving simplicity of construction with the maximum of repetition. Speed, simplicity, and economy are of great concern to the man who eventually pays for the completed construction work. Economy interests him as much as safety; both are fundamental to good design.

These are the points a good designer has always to give his attention to; and this is why he will invariably make use of design charts or tables wherever possible, so as to have as much of his time as he can available for watching the over-all matters of conception and economy.

5

Bending moments, T-beams and slabs

Bending moments

We concluded the previous chapter with our first example of reinforced concrete design: it was for a single-span beam. And we saw that before we could tell what resistance moment we should design the beam for, we needed to know what bending moment was being caused by the loads on the beam; then the beam was designed so that its resistance moment would cater for the worst possible bending moment. Our case was of a simply supported beam carrying a uniform load, for which we took the bending moment from Fig. 5.1(d) as

$$M = \frac{UL}{8}.$$

Now we need to understand how to derive other bending moments, and what information we get from examining bending moment diagrams.

A bending moment is always a force multiplied by a perpendicular distance. The simplest case is a cantilever of length L carrying a point load P at its unsupported end (see Fig. 5.1(a)). By visualizing the curved shape of the cantilever when loaded, it is clear that the tension in this case will be at the top of the member, which is the opposite of what we had with our simple beam resting on two supports. Hence we regard our cantilever moment as *negative*. Its value, at the position X, is $-P \times x$: and is a maximum at the point of support, where

$$M = -PL.$$

When a cantilever carries a uniformly spread load U, the centre of this load clearly acts at a point half-way along the length, so that the maximum moment at the point of support is

$$M = -U \times \frac{L}{2}$$

$$= -\frac{UL}{2} \text{ (see Fig. 5.1(b))}.$$

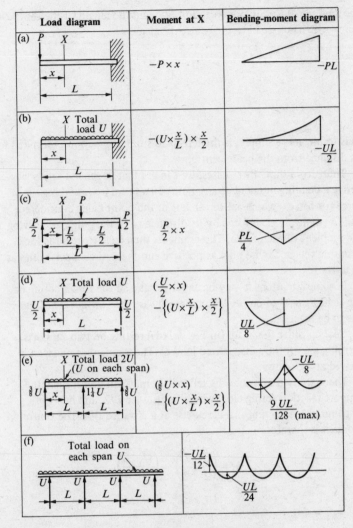

Load diagram	Moment at X	Bending-moment diagram
(a)	$-P \times x$	$-PL$
(b)	$-(U \times \frac{x}{L}) \times \frac{x}{2}$	$-\frac{UL}{2}$
(c)	$\frac{P}{2} \times x$	$\frac{PL}{4}$
(d)	$(\frac{U}{2} \times x) - \left\{ (U \times \frac{x}{L}) \times \frac{x}{2} \right\}$	$\frac{UL}{8}$
(e)	$(\frac{3}{8}U \times x) - \left\{ (U \times \frac{x}{L}) \times \frac{x}{2} \right\}$	$-\frac{UL}{8}$ $\frac{9\,UL}{128}$ (max)
(f)		$-\frac{UL}{12}$ $\frac{UL}{24}$

FIG. 5.1. Typical bending-moment diagrams.

With the beam in Fig. 5.1(c) resting on two supports, carrying a central point load P, the maximum bending moment will be at the centre. It will be noticed that if the beam and its load and support reactions are all inverted, this case is exactly similar to two cantilevers (as in Fig. 5.1(a)) put back to back, the load on the 'cantilever' being the amount of the support reaction $P/2$, and the length of the cantilever being $L/2$. Thus we have the centre moment

$$M = \frac{P}{2} \times \frac{L}{2}$$

$$= \frac{PL}{4}.$$

This result, more simply, is the left-hand support reaction multiplied by its distance from the beam centre.

Thus, from these three cases, we can see that in general terms we derive a bending moment at any point on a beam by (a) taking the force (or forces) which act to the left of the point being considered, and (b) multiply each force by its distance from the point, so arriving at the moment of each force. The moments then have to be added up, counting the clockwise ones as positive and the anti-clockwise ones as negative.

In some situations it may be easier to take all the forces acting on the right instead of on the left, in which case the convention of signs has to be reversed.

Now consider the beam (in Fig. 5.1(d)) resting on two supports, carrying a uniformly distributed load U. The moment at midspan is arrived at as follows.

There are two forces to the left of our midspan point, the left support reaction acting clockwise at a distance $L/2$, and half the distributed load acting anti-clockwise at a distance $L/4$. The centre moment is therefore

$$M = \left(\frac{U}{2} \times \frac{L}{2} \right) - \left(\frac{U}{2} \times \frac{L}{4} \right)$$

$$= \frac{UL}{8}.$$

This is the expression we used in the previous chapter.

In the foregoing we have only worked out the moments at the positions where these are a maximum. The reader will recognize that the centre column of Fig. 5.1 gives the general formulae for the moment at position X in each case: from these he can calculate numerical values for different points along each beam, and if these are plotted like points on a graph using for a base the line of the beam, it will be found that the bending moments vary from point to point in accordance with the diagrams given in the right-hand column of Fig. 5.1.

In the cases of the cantilevers at (a) and (b), supported at one end only, effective stability is achieved by the support ends being held rigidly in direction (*encastré*). The beams at (c) and (d) are stable merely resting on supports at both ends without the need for being held in direction: such beams are known as *simply supported*. The beams at (e) and (f) are sets of *continuous beams* where each beam is supported at both ends, but in addition the ends when the beam continues into the adjoining span are restrained in direction to varying degrees depending on the balance of the loads on either side of the intermediate supports.

It is more difficult to derive the bending moment diagrams for continuous beams; for example in case (e) it is not so obvious how one should determine the amounts of the support reactions. The detail of this is beyond the scope of this book; but when suitable bending-moment diagrams are made available, the reader will want to know how to read these and what they mean. They are crucial to any understanding of where the reinforcements in the members have to be put.

Firstly the bending-moment diagram gives a guide as to the deflected shape of the beam. Where the bending moment is negative, the beam bends into a curved shape with the inside of the curve on the underside; the commonly used term for this is a *hogging moment*, and beam types (a), (b), and the central part of (e) are examples. Conversely, where the bending moment is positive, the beam bends with the inside of the curve at the top; and this, naturally enough, is known as a *sagging moment*, and examples of this are beam types (c), (d), and the outer parts of (e).

The deflected shape of (e), suitably exaggerated, is shown in Fig. 5.2. The change-over points between the parts of the beam subject to sagging moments and hogging moments are known as the *points of contraflexure*; these correspond precisely with the points of zero moment on the bending-moment diagram. As we always provide the main reinforcements in the tensile zones of the beam, so as to make up for the weakness of the concrete in tension, it follows that the portions of beam subject to hogging moments have their tensile steel in the top and those subject to sagging moments have their tensile steel in the bottom.

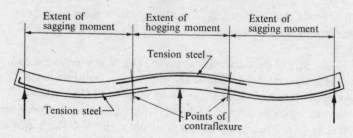

FIG. 5.2. Bending of a 2-span continuous beam (exaggerated).

By comparing the bending-moment diagrams (d) and (e) a popular misconception can be quickly dispelled. It is common experience that the effect of continuity over the centre support in (e) gives a stiffer arrangement than if the double span were covered by two separate beams both simply supported as in (d); in other words, for equal loads and spans, the deflections in case (e) will be less than in case (d). However this fact is sometimes misconstrued as indicating that the two-span continuous beam (assumed to be of equal strength) would be able to carry a greater load than two separate simply supported beams: but by looking at the two bending moment diagrams we see the maximum moment in each case is equal to $UL/8$, and in case (e) has merely been moved from midspan to the position of the central support. It is only in situations of continuity over many spans, as in (f), that we get the real benefit of reduced moments. When we come to T-beams, later in this chapter, we shall find that sagging moments can be accommodated more efficiently than hogging moments, so we see the two-span continuous beam at (e) is not as attractive as is sometimes imagined.

In practice we consider the loads which act on our beams as coming into either of the categories of dead loads or imposed loads. Dead loads are permanent and constant, and include the self-weight of the structure itself together with the weights of walls, permanent partitions, and all building finishes such as ceilings, plaster, and flooring. Imposed loads are those which may or may not be present at any time, such as people, stores, machinery, and vehicles on floors, and snow, wind, flooding, and people on roofs.

The reason for distinguishing between the two, is that the effects of imposed loading can sometimes be more onerous when some of the load is *removed*, leading to assymetry or out-of-balance. The beam arrangements shown in Fig. 5.3 are identical to those we had in Fig. 5.1

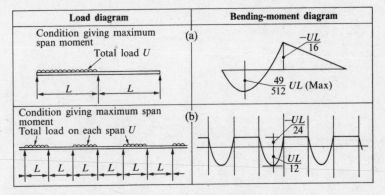

FIG. 5.3. Bending-moment diagrams: imposed-loading on some spans only.

(e) and (f), except that parts of the loading have been removed; and if we compare the bending moment diagrams, we see in both cases that the removal of part of the load has had the effect of increasing the bending moments in other parts of the line.

For an important beam, with varying spans and imposed loads, many different load combinations may have to be considered before the absolute maximum bending moment at every point along the beam can be determined. Before the days of computers this used to be a very tedious process for the designer; but now these calculations can be readily handled by normal computer programs, and the worst bending moments drawn out quickly and automatically by the use of a computer-linked data plotter.

When the bending moment diagrams for all the possible variations of imposed loading on a continuous beam system have been plotted, the line is then drawn enclosing the worst of these moment diagrams. To this are then added the moments due to the dead loads; and the curve of the resultant maximum moments is known as the *envelope diagram* for which the beam will have to be designed. Fig. 5.4 shows an example of this procedure for a two-span system as we had in Figs. 5.1(e) and 5.3(a); the dead loads are permanent and can act in one configuration only whereas there are three configurations possible for the application of the imposed loads. The envelope diagram gathers together the worst of all the conditions for this simple case.

Where there are more than two spans the number of possible configurations becomes greater and more complicated. On normal jobs it is then usual to short-cut the procedure and use Table 5.1 to arrive at

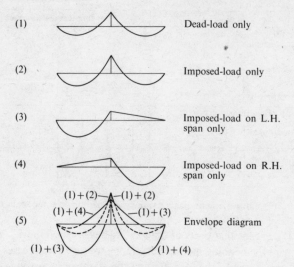

FIG. 5.4. Bending-moment envelope diagram for a 2-span continuous beam.

approximate maximum bending moments at the various points along the beam. The conditions governing the use of this Table are

   (i)   there must be at least three spans;
  (ii)   the shortest span must not be less than 85 per cent of the longest;
 (iii)   the imposed load must not exceed the dead load.

T-beams and L-beams

The beam which we considered in Chapter 4 had a rectangular cross-section. Some reinforced-concrete structures do have a number of such beams, but in building frames it is common to arrange for the floor slabs to be cast monolithically with the beam ribs. Slabs formed in this way can be considered as part of the beam, and contribute

TABLE 5.1
Ultimate bending moments for the design of continuous beams

At outer support	Near middle of end span	At first interior support	At middle of interior spans	At interior supports
0	$\dfrac{UL}{11}$	$-\dfrac{UL}{9}$	$\dfrac{UL}{14}$	$-\dfrac{UL}{10}$

appreciably to its strength; the slab thus forms a top flange to the beam. For an internal beam, the slab is available on both sides of the beam, and this is then known as a T-beam. At an external beam, where the slab occurs on one side only, this is known as an L-beam.

The effect of the flange is to increase the width of concrete available to resist compression where we have a sagging moment. Where a hogging moment exists and the upper part of the beam is in tension, the same advantage of course does not exist, and the flange has to be disregarded in any strength calculation.

Some limit has to be set on the width of slab that can be included as part of the beam and CP.110 (1972) recommends the following.

For T-beams, the lesser of

(i)   the actual width of the flange,
(ii) (a)   for single-span beams: the width of the web of the beam plus 0·2 times the span,
      (b)   for continuous beams: the width of the web of the beam plus 0·14 times the span.

For L-beams, the lesser of

(i)   the actual width of the flange,
(ii) (a)   for single-span beams: the width of the web of the beam plus 0·1 times the span,
      (b)   for continuous beams: the width of the web of the beam plus 0·07 times the span.

A measure of the relative advantage of a T-beam over the equivalent rectangular beam can be seen by comparing the resistance moments of the range of typical beams given in Table 5.2. Taking the 600 x 250 beam as a random example, the resistance moment of the T-beam is

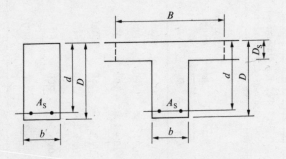

TABLE 5.2
Properties of typical beams

Physical dimensions (see Fig. on previous page)				Ultimate resistance moment (M_u) of concrete, and corresponding tensile steel requirements						Moment of inertia of concrete section	
				Rectangular beam			Tee-beam			Rect-angular beam section	Tee-beam section
Beam size $D \times b$	Slab depth D_s	Assumed flange width B	Effective depth d	Concrete res. mom. (eqn 4.6)	Steel reqd. A_s	Bars provided No x dia	Concrete res. mom. (eqn 5.1)	Steel reqd. A_s	Bars provided No x dia		
(mm x mm)	(mm)	(mm)	(mm)	(kN m)	(mm²)		(kN m)	(mm²)		(mm⁴ x 10⁹)	(mm⁴ x 10⁹)
400 x 150	100	800	350	69	709	4–16	240	2163	2–25 / 2–32	0·80	1·57
500 x 200	100	1000	440	145	1188	4–20	390	2704	4–32	2·08	3·91
600 x 250	150	1200	530	263	1791	6–20	819	4869	6–32	4·50	8·37
700 x 250	150	1400	620	360	2096	2–25 / 4–20	1144	5678	2–25 / 6–32	7·15	13·9
800 x 300	200	1600	710	567	2883	6–25	1952	8654	2–32 / 6–40	12·8	24·9
900 x 300	200	1800	800	720	3249	2–32 / 4–25	2520	9736	8–40	18·2	36·8

Grade 25 concrete, and 425 steel

819/263, i.e. more than 3 times that of the rectangular beam. Furthermore the T-beam, by concentrating the compression force in the flange, as far away as possible from the tension steel, develops an increased lever arm; so we see that the 3-times increased resistance moment is achieved by using only 4869/1791, i.e. 2¾ times as much steel. An L-beam shows similar advantages, but to a lesser degree.

In Table 5.2 the concrete resistance moment and the steel area for each of the rectangular beams have been calculated using the Design Chart given in Fig. 4.7. The corresponding calculations for the T-beams are very straightforward, as follows.

Provided the flange thickness D_s is not greater than ½d (as applies in all these cases), the whole of the flange can be relied upon to take its share of the compression force. Thus the ultimate resistance moment from consideration of the concrete is

concrete stress x flange area x lever arm,

i.e. $$M_u = (0.4 f_{cu}) \times (B \times D_s) \times \left(d - \frac{D_s}{2}\right),\qquad(5.1)$$

and from consideration of the steel the ultimate resistance moment is

steel stress x steel area x lever arm,

i.e. $$M_u = (0.87 f_y) \times A_s \times \left(d - \frac{D_s}{2}\right).\qquad(5.2)$$

The reader should remember that T-beams and L-beams are both subject to the important limitation that the usefulness of the flange only applies where it is in compression, i.e. when the beam is resisting a sagging moment. With hogging moments no advantage is gained; and since we see from Table 5.1 that hogging moments are generally greater than sagging ones, this is clearly a point that needs careful watching. Fortunately, the worst peaks of the negative moment diagrams are very steep, so that by the time we reach the face of the supporting column the moment is already considerably less than at its maximum; and if the rectangular concrete section is then still in difficulty its strength can be augmented locally by the addition of some compression steel in the bottom of the beam.

One further point covered by Table 5.2 is worth drawing attention to. This is the end two columns which give *moments of inertia* for rectangular and T-beams. Up to the present we have been concerning ourselves only with the stresses in beams; but later, when we are dealing with columns in Chapter 7, we shall find ourselves involved with moments of inertia of different sections when considering how the bending moment from a beam is distributed when the beam frames into a column. The moment of inertia of a simple rectangle can be calculated relatively easily from the formula

$$I = \frac{bD^3}{12},$$

where b is the breadth, and D is the depth; but for T-beam sections the calculation is not so simple. The moments of inertia of the T-beams in Table 5.2 can be seen to be about double those of the corresponding rectangular beams; this is quite normal, but of course depends on the dimensions chosen.

T-beam example

Let us now design a run of T-beams continuous over three equal spans of 7·0 m as indicated in Fig. 5.5(a). This beam run is parallel to other similar beam runs in a building which are spaced 3·0 m apart on both sides. The slab forming the flange of the beam is 150 mm thick. The total characteristic loads, both taken here as uniformly distributed, are 100 kN dead-load, and 80 kN imposed-load. Using 1·5 as the partial safety-factor for loads (γ_f), we have

$$\text{design load} = 1\cdot5 \ (100 + 80)$$
$$= 270 \text{ kN}.$$

So from Table 5.1 we have

support moments:

$$M_A = M_D = 0$$

$$M_B = M_C = -\frac{270 \times 7}{9} = -210 \text{ kN m},$$

span moments:

$$M_{AB} = M_{CD} = \frac{270 \times 7}{11} = 172 \text{ kN m}$$

$$M_{BC} = \frac{270 \times 7}{14} = 135 \text{ kN m}.$$

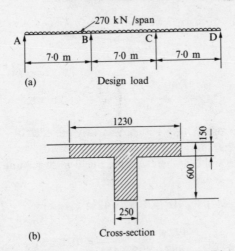

(a) Design load

(b) Cross-section

FIG. 5.5. 3-span continuous reinforced-concrete T-beam.

Referrring to Table 5.2 we see that a rectangular beam rib
500 mm x 200 mm would be inadequate to resist the support moment
of −210 kN m, so we select the next largest size of 600 mm x 250 mm
which is good for 263 kN m and therefore suitable.

To determine the width of flange available to act with the beam rib,
we have, from the limit conditions given earlier in the chapter, that this
is to be the lesser of

(i)   the actual flange width = 3000 mm
(ii) 250 mm + (0·14 x 7000 m) = 1230 mm.

1230 mm is slightly more than the width of 1200 mm allowed for in
Table 5.2, which we see could take a span moment of 819 kN; so
without needing to use equation (5.1) we know that the span moment
of 172 kN in our own case is amply catered for. The cross-section of
our beam is therefore satisfactory as shown at Fig. 5.5(b).

Allowing for two layers of steel bars plus a cover of concrete to protect
them, we arrive at our effective depth (d) as 600 − 70 = 530 mm.

Then at supports B and C, where our section is rectangular, we have

$$\frac{M}{bd^2} = \frac{210 \times 10^3 \times 10^3 \text{ N mm}}{250 \times 530^2 \text{ mm}^3}$$

$$= 3 \cdot 00 \text{ N/mm}^2.$$

From the Chart of Fig. 4.7 we have

$$\frac{100\,A_s}{bd} = 0.97,$$

giving
$$A_s = \frac{0.97 \times 250 \times 530}{100} = 1280 \text{ mm}^2,$$

which we see from Table 4.1 is conveniently provided by 4–20 mm bars (area = 1260 mm^2).

In spans **AB** and **CD** where we have the T-beam section, from equation (5.2) we have

$$172 \times 10^3 \times 10^3 = (0.87 \times 425) \times A_s \times (530 - 75),$$

giving
$$A_s = \frac{172 \times 10^3 \times 10^3}{0.87 \times 425 \times 455} = 1020 \text{ mm}^2,$$

which is provided by 2–20 mm and 2–16 mm bars (area = 1030 mm^2).

Similarly in span **BC**

$$A_s = \frac{135 \times 10^3 \times 10^3}{0.87 \times 425 \times 455} = 805 \text{ mm}^2,$$

which is provided by 4–16 mm bars (area = 804 mm^2).

Slabs

Reinforced concrete slabs are a normal form of construction for the floors and roofs of most modern buildings. The simplest example would be a slab of one span supported on two parallel walls. Such a slab would span in one direction only, and would have a bending-moment diagram as in Fig. 5.1(d), and can be designed the same way as we designed our beam in Chapter 4, treating the slab as though it were a wide shallow beam.

One-way spanning slabs can equally well be made continuous over supports. They can also be built as cantilevers. The various bending-moment diagrams developed earlier in this chapter apply. The following thicknesses are a useful rule-of-thumb guide for such slabs:

> simply supported (single span) : 1/20th span;
> continuous at both supports   : 1/25th span;
> cantilever slab   : 1/8th span.

As an example of a simply supported slab spanning one way, consider a slab required to carry a characteristic applied load (dead + imposed) of $5 \cdot 0$ kN/m^2 over a span of $3 \cdot 0$ m measured between centres of supports. From the simple rule given above, a convenient slab thickness will be $1/20 \times 3000 = 150$ mm. The total characteristic load intensity is then made up of

$$\text{self weight of slab} : 0 \cdot 15 \text{ m} \times 24 \text{ kN/m}^3 = 3 \cdot 6 \text{ kN/m}^2$$

$$\text{applied load} \qquad : \qquad \qquad 5 \cdot 0 \text{ kN/m}^2$$

$$\overline{8 \cdot 6 \text{ kN/m}^2}$$

Design load intensity is then $8 \cdot 6 \times 1 \cdot 5 = 12 \cdot 9$ kN/m^2.
Considering one metre width of slab,
the total design load $U = 12 \cdot 9 \times 3 \cdot 0 = 38 \cdot 7$ kN.

Therefore the midspan moment $M = \dfrac{UL}{8}$

$$= \frac{38 \cdot 7 \times 3 \cdot 0}{8}$$

$$= 14 \cdot 5 \text{ kN m.}$$

Taking the effective depth (d) of our 150 mm slab as 120 mm, we have

$$\frac{M}{bd^2} = \frac{14.5 \times 10^3 \times 10^3}{1000 \times 120^2} = 1 \cdot 01 \text{ N/mm}^2.$$

From the Chart at Fig. 4.7 we read

$$\frac{100 \, A_s}{bd} = 0 \cdot 28,$$

whence $\qquad A_s = \dfrac{0 \cdot 28 \times 1000 \times 120}{100} = 336 \text{ mm}^2.$

By referring to Table 5.3 for areas of steel bars per metre width of slab, we see this is conveniently provided by 12 mm dia bars at 300 mm crs (area = 377 mm^2).

These bars are known as the *main reinforcement*, and provide the necessary strength for the slab to span from support to support; however to distribute loads in the other direction, and as a precaution against cracking, the slab also needs *secondary reinforcement* at

TABLE 5.3
Areas of steel bars (mm^2) per metre width

Bar dia (mm)	Spacing of bars (mm)								
	50	75	100	125	150	175	200	250	300
6	566	377	283	226	189	162	142	113	94·3
8	1010	671	503	402	335	287	252	201	168
10	1570	1050	785	628	523	449	393	314	262
12	2260	1510	1130	905	754	646	566	452	377
16	4020	2680	2010	1610	1340	1150	1010	804	670
20	6280	4190	3140	2510	2090	1800	1570	1260	1050
25	9820	6550	4910	3930	3270	2810	2450	1960	1640
32	16100	10700	8040	6430	5360	4600	4020	3220	2680
40	25100	16800	12600	10100	8380	7180	6280	5030	4190

right-angles to the span not less than 0·12 per cent of the gross concrete area. Thus we have secondary steel

$$= \frac{0 \cdot 12 \times 1000 \times 150}{100} = 180 \text{ mm}^2,$$

which is provided by 10 mm dia bars at 400 mm centres (area = 196 mm²). See Fig. 5.6.

If the slab, carrying the same load, had been *continuous* over a number of such spans, all equal, as often occurs in practice, we could have chosen a lesser thickness and adopted a reduced bending moment. Referring again to our rule for thicknesses, we see a suitable slab would now be 1/25 x 3000 = 120 mm.

The total characteristic load intensity is then:

self weight of slab : 0·12 x 24 kN/m³ = 2·9 kN/mm²

applied load   :   5·0 kN/mm²

                                          7·9 kN/mm²

Design load intensity is then 7·9 x 1·5 = 11·8 kN/mm².
Total design load per metre U = 11·8 x 3·0 = 35·4 kN.

From Figs. 5.1(f) and 5.3(b) we see that the maximum support moments occur when all spans are fully loaded and are equal to $-(UL/12)$, and the maximum midspan moments occur when alternate spans are unloaded and are equal to $+(UL/12)$. Actually in practice the midspan moment cannot be as great as $+(UL/12)$ because even with the applied load removed from alternate spans, the self-weight of the slab

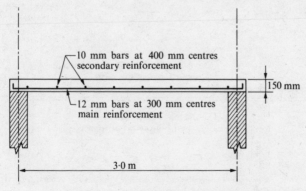

FIG. 5.6. 1-way spanning slab (simply supported).

will continue to act; thus the case illustrated in Fig. 5.3(b) is extreme, and taking a midspan moment of $+(UL/12)$ gives a hidden margin of security. Nevertheless the use of equal numerical values of $UL/12$ for support and midspan moments is very convenient and often adopted.

Thus we have
$$M = \frac{35 \cdot 4 \times 3 \cdot 0}{12}$$

$$= 8 \cdot 8 \text{ kN m.}$$

Taking d of our 120 mm slab as 90 mm,

$$\frac{M}{bd^2} = \frac{8 \cdot 8 \times 10^3 \times 10^3}{1000 \times 90^2} = 1 \cdot 08 \text{ N/mm}^2 .$$

From the Chart in Fig. 4.7 we read

$$\frac{100 A_s}{bd} = 0 \cdot 29,$$

whence
$$A_s = \frac{0 \cdot 29 \times 1000 \times 90}{100} = 260 \text{ mm}^2 .$$

This is provided by 10 mm bars at 300 mm crs (area = 262 mm^2). See Fig. 5.7.

Comparing this result with the example before for a simply supported slab, we see the considerable economy achieved by designing the slab to be continuous. Not only is the continuous slab 30 mm thinner, and therefore lighter; it is also more economical of concrete materials and needs lighter reinforcing steel. This is common experience.

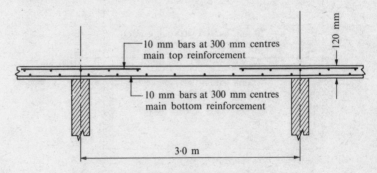

FIG. 5.7. 1-way spanning slab (continuous).

Where slabs are used to form the flanges of T-beams, they lend themselves naturally to such continuous treatment. Indeed top-steel in such slabs across the tops of T-beam ribs is a necessary requisite for the slab to act effectively as a flange.

In all that we have discussed about slabs so far, we have been considering the slab as supported on two opposite sides only. However it is possible there might be beams (or walls) arranged so as to support the slab on four sides instead of only two; and if the four beams were equal distances apart in both directions, so that the slab they supported was a square, it would be as easy for the slab to span from north to south as it would from east to west. Indeed if adequate reinforcements were provided in both directions this is exactly what would happen, and the moments both ways would be only half the values we have taken previously for one-way spanning. As a consequence of this, the areas of steel required each way are also halved. Care then needs to be taken that there is not less than a minimum percentage of steel for practical considerations; and for strength purposes a suitable minimum for *main reinforcement* in slabs is 0·15% *bd* when using high yield bars, or 0·25% *bd* when using mild steel bars. In slabs spanning two ways, clearly the steel in *both* directions is needed for strength, and therefore classed as main reinforcement.

Where the beams (or walls) supporting two-way spanning slabs are not equidistant in both directions (i.e. where the north to south and east to west spans are different), the moments clearly cannot be taken on the simple 50–50 basis referred to above. The division then becomes much more complicated, as the reader will find by reference to the equations and tables given in CP.110 (1972).

Fig. 5.8 illustrates a *ribbed slab* and a *waffle slab*. These are refinements of the normal solid slab, and have parts of the concrete omitted on the underside so as to save weight. This is clearly admissible in the parts of the span where the sagging moments occur, since the only purpose this concrete serves below the neutral axis is to hold the tension steel in place below the compression zone, and to resist shear. With the voids arranged so as to come between the bar positions no harm results, and we get the benefit of less weight to carry on our structure. In effect, ribbed slabs are essentially a lot of little T-beams of slab dimension side by side: and as we saw earlier in this chapter, it is perfectly satisfactory for the voids to extend up even above the level of the neutral axis, provided sufficient concrete remains at the top to take the compression force.

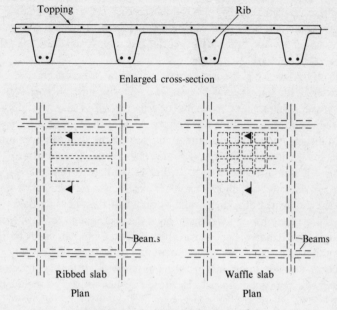

FIG. 5.8. Typical ribbed and waffle slabs.

The ribbed slab illustrated spans in one direction only; waffle slabs can be used to act as two-way spanning. The voids in these slabs can be formed by laying clay tiles or lightweight blocks on the soffite formwork, and concreting these in to form part of the finished construction. Alternatively the voids can be formed using temporary thin polypropylene formers which are subsequently removed from the underside, and can be re-used many times. The main reinforcements in these types of slabs are normally two bars in the bottoms of each rib; and in some cases, depending on the dimensions and other considerations, a mesh of smaller bars is also provided in the top concrete as a precaution against cracking and local fracture.

This section of the chapter would be incomplete without a mention of flat-slab construction. A typical flat-slab arrangement is indicated at Fig. 5.9. This form of construction has the feature of providing large open areas without the need for downstand beams, the whole of the support coming directly from columns. Such an arrangement has great merit in buildings where air-conditioning services need to be run unimpeded in the spaces above false ceilings; also in store-rooms where forklift trucks can make use of the full floor-height and enjoy freedom

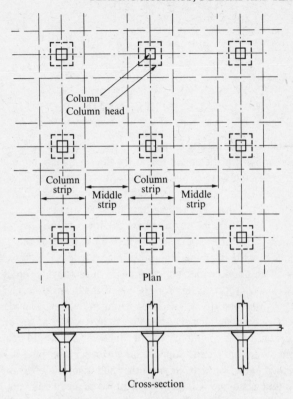

FIG. 5.9. Typical flat-slab construction.

of circulation; also in circumstances where natural ventilation of large spaces is sought by encouragement of free through-draughts.

Although flat-slab construction gives the appearance of having no beams, these are incorporated in essence as *column strips* within the slab thickness; the remainder of the slab, the *middle strips*, then act as two-way slabs, spanning onto the column strips. The details of calculation lend themselves to considerable refinement.

The weakest feature of flat-slab construction is the high concentration of shear stress that occurs at the columns. For this reason the heads of the columns are normally splayed: even so it is generally necessary also to make the slab itself thicker than would be usual in ordinary beam-and-slab construction. The greater weight of the thicker slab can then to some extent be avoided, by waffling the undersides of the sagging-moment areas in the manner discussed earlier.

6

Shear strength

A beam can be made strong enough to resist the effects of *bending*,
being big enough for the concrete to take the compression, and having
enough reinforcing steel to take the tension; yet this same beam can fail
as we get near to the supports as a result of a completely different
action known as *shear*. The total external force acting across the
longitudinal axis of the beam anywhere is known as the *shear force*.
The *shear strength* of the beam is provided by the part of the beam
which connects the concrete compression area to the steel tension area.

Consider an unreinforced concrete beam of rectangular cross-section
resting on two supports. The proportions of the beam are such that it
can carry a total load of $2V$† without failing in bending at midspan.
Suppose now the beam has diagonal cracks in it at $45°$ as shown in Fig.
6.1(a). Evidently the middle part with the load on it would drop away
and complete failure occur. Suppose now we stick the three parts
together, with glue, cement, or any other adhesive, it is clear the beam
will be sound again and will carry such a load as just breaks the joint
again. Clearly the stress in Fig. 6.1(a) across the joints is a tensile stress,
as the middle pulls away from the ends. In practice the beam is not
broken, but will break if the tensile strength of the concrete across
these planes is not sufficient. This is how failure by shear in concrete
beams nearly always occurs.

Consider now Fig. 6.1(b). Suppose here the beam is divided by
diagonal planes at right angles to those previously referred to. It is clear
that the centre part is pressed by the load against the end parts, and
that the stress at the joints is one of compression. The beam will
successfully carry its load if the centre part is carefully fitted to the two
ends, without any glue or adhesive whatever.

We see, then, that the ends of a beam are in a condition of tension
across diagonal planes in one direction, and of compression across
planes at right angles, all as indicated at Fig. 6.1(c). The beam will

---

† V is the symbol used in CP.110 (1972) for shear force due to ultimate loads.

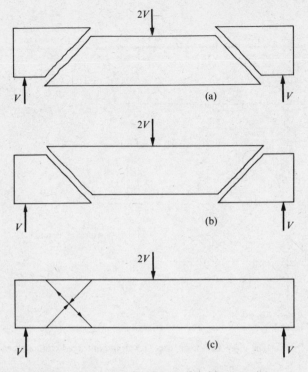

FIG. 6.1. Shear in a concrete beam: the basic principles.

break if the tension or compression become greater than the concrete can stand; and as we have already explained that concrete is about ten times as strong in compression as in tension, clearly the tension failure shown at Fig. 6.1(a) will occur first and is the chief danger to be guarded against.

Let us now evaluate the effect of a shear force applied to our unreinforced concrete beam. Experiments show that failures due to shear invariably occur roughly at 45°, so we shall confine our attention to such cases.

Fig. 6.2(a) shows the tensile stress of t N/mm² acting across our 45° *diagonal plane* near the end of the beam. Compare this with Fig. 6.1. So long as the stress t does not exceed the tensile strength of the concrete, no crack will form, and the beam will hold up.

If we take b for the breadth of the beam, and d for the effective depth, then the length of the diagonal plane is $\sqrt{2}\,d$, and the area of the

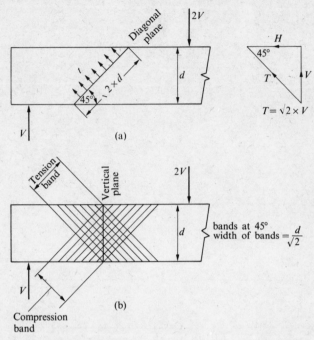

FIG. 6.2. The effects of shear forces on an unreinforced concrete beam.

plane is $\sqrt{2}\ bd$. Therefore the total tension force across the plane is

$$T = \sqrt{2}\ bdt.$$

This diagonal force can be resolved into its vertical and horizontal components by use of the triangle of forces as shown at the right-hand diagram of Fig. 6.2(a). V, the vertical component, is what has to prevent the centre part of the beam from dropping down†; in other words it is the *shear strength of the beam* and has to match the shear force applied by the external loading.

Thus
$$V = \frac{T}{\sqrt{2}} = \frac{\sqrt{2}\ bdt}{\sqrt{2}}.$$

So
$$V = bdt. \qquad (6.1)$$

---

†H, the horizontal component, is what builds up to produce the horizontal tension in the bottom of the beam, and the horizontal compression in the top of the beam.

This means that the shear strength of the beam is equal to its cross-sectional area multiplied by the tensile stress in the concrete.

This is a very important result, and so simple it is easy to misunderstand. It relates a *shear force V* acting at right angles to the axis of the beam, to a *tensile stress t* on a diagonal plane at 45°; yet the area on which we calculate the stress is measured not diagonally but at right angles to the beam, that is to say parallel to the shear force. In other words

$$t \text{ (an inclined tensile stress)} = \frac{V}{bd}.$$

This expression V/bd, the shear force divided by the cross-sectional area, is commonly called the 'shear stress', but really means nothing because the beam fails by diagonal tension as already explained. To aggravate the risk of this being misunderstood, equation (6.1) is more normally written

$$V = bdv$$

or

$$v = \frac{V}{bd} \tag{6.2}$$

where v is actually defined as the shear stress.

Equation (6.2) is as given in CP.110 (1972); and to ensure we do not get muddled over it, it will be worth deriving the formula a second time in a slightly different way.

Consider now the *vertical plane* within the beam as shown in Fig. 6.2(b), and consider the forces acting across this plane. The vertical components of these forces must match the applied vertical shear force at the section. There are an infinite number of lines at 45° which can be drawn in the beam to represent tension forces, but we are only concerned here with those passing through the one vertical plane drawn. These give a tension band whose width is less than the depth of the beam and actually $d/\sqrt{2}$. Hence the total tension crossing the vertical plane is $bdt/\sqrt{2}$, and the vertical component of this is

$$T_v = \frac{1}{\sqrt{2}} \times \frac{bdt}{\sqrt{2}} = \frac{bdt}{2}.$$

But there is also a compression band which crosses the same vertical plane; and from Fig. 6.1 we can see that the compression force is the same amount as the tension, and as it acts on an equal area, the compression stress is equal to the tension stress, and the vertical component

$$C_v = \frac{bdt}{2} \quad \text{as before.}$$

Hence the total shear strength of the beam

$$V = T_v + C_v = \frac{bdt}{2} + \frac{bdt}{2};$$

so $\qquad\qquad V = bdt,$

as we arrived at at equation (6.1), which leads on to the same result as at equation (6.2). This method of taking the vertical component of all forces acting across a vertical plane is particularly useful when considering complicated reinforced beams; also by bringing the contribution of the diagonal compression into mind, it may make easier the understanding of equation (6.2).

One way of preventing diagonal tension cracks in a beam would obviously be to have steel bars bent-up at right-angles to where these cracks are expected to come; and indeed this is sometimes done, and is discussed later in this chapter. However longitudinal reinforcements have to be provided in the beam in any case to give the necessary strength for resisting bending; and it is more normal to resist shear by making use of these acting in conjunction with vertical links as shown in Fig. 6.3. Indeed because of the risk of accidental cracking of concrete due to shrinkage and temperature effects, it is wise never to have less than about 0·2 per cent longitudinal tension reinforcement and about 0·2 per cent vertical links†. It is also important that the spacing of the links does not exceed 0·75 times the effective depth of the beam; it is then clear from Fig. 6.3 that it would be impossible for a diagonal crack at 45° to form anywhere without first pulling on all the longitudinal bars and at least one of the vertical links.

Longitudinal steel

Owing to the untrustworthy performance of concrete in tension, even its limited tensile strength as implicit in equation (6.2) can be relied

---

† Longitudinal reinforcement should be not less than 0·15 per cent when using high yield reinforcement, and not less than 0·25 per cent when using mild steel reinforcement. For links the corresponding figures are 0·12 per cent and 0·20 per cent.

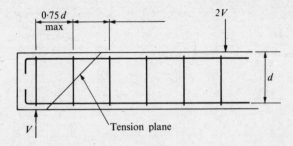

FIG. 6.3. Shear forces resisted by the combined action of the longitudinal reinforcement and links.

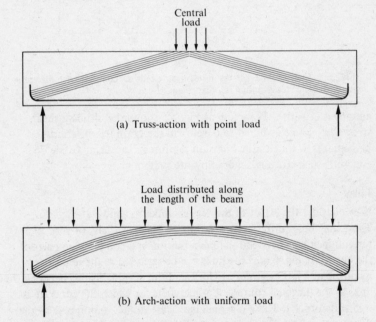

(a) Truss-action with point load

(b) Arch-action with uniform load

FIG. 6.4. Shear forces resisted by truss or arch action.

upon only in relation to the percentage of longitudinal steel serving to hold the beam together. This steel assists the beam in forming a kind of arch or truss as indicated in Fig. 6.4, the concrete forming the inclined compression members; and the strength of such an arch naturally depends on the quantity of reinforcement in the bottom of the beam preventing the arch from spreading. It is essential of course that the steel bars are well secured at the ends of the beam to ensure that the

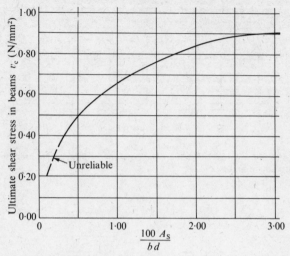

FIG. 6.5. Ultimate shear stress in beams of Grade 25 concrete with different percentages of longitudinal reinforcement.

arching is effective. The graph at Fig. 6.5 shows the ultimate shear stress that can be assumed to develop in the beam for different percentages of longitudinal tension reinforcement, using Grade 25 concrete; such stress is denoted by the symbol v_c.

Links

Now consider the effect of the vertical links, as shown at Fig. 6.6. These may be of different forms, but consist essentially of vertical tension members between the main tension and compression parts of the beam. At Fig. 6.6(a) the links can be regarded as the tension members of a lattice girder or truss, and the diagonal bands between the links as the diagonal compression members for which the concrete is well suited to act. These diagonal bands are similar in principle to the direct compression bands shown at Fig. 6.4, except that being steeper they are more efficient in providing strength in terms of vertical shear. The links actually augment the strength of the direct arching effect, by catching hold of the arch and hitching it up into the truss system shown in Fig. 6.6(a). The greater the provision of links, the greater the proportion of the shear force that gets carried by the more efficiently disposed compression bands of the link truss system.

If the links are spaced apart (s_v) the same amount as the effective depth of the beam (d), it would clearly be just possible for a 45°

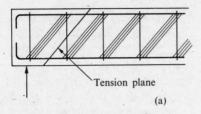

Cross-section

(a)

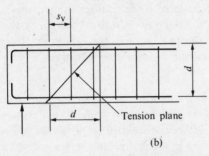

(b)

FIG. 6.6. Vertical links.

diagonal crack to form between adjoining links, so the beam could fail
in shear and the links be completely ineffective; this is why links must
never be further apart than $0 \cdot 75\, d$. On the other hand, if they are much
closer, any plane at $45°$ will cut more than one, and the strength will be
proportionately greater. If

A_{sv} is the cross-sectional area of the two legs of a link,
s_v is the spacing of the links,
f_{yv} is the characteristic strength of the link steel,

then the tension force in a link at failure is $A_{sv}f_{yv}$. Now the number of
links crossing any diagonal plane will be d/s_v (see Fig. 6.6(b)), so the
resistance to shear provided by the links at failure will be $A_{sv}f_{yv}(d/s_v)$;
and if we apply our partial safety factor to the steel, we have the
ultimate shear strength provided by the links as

$$\frac{A_{sv}\,(0 \cdot 87\, f_{yv})\, d}{s_v}. \tag{6.3}$$

Note that this is a strength arising from the links setting up a
completely different internal mechanism of behaviour from that which
occurs when the concrete is resisting its natural tendency to fail by

diagonal tension. The effect of equation (6.3) can therefore be added to the effect of v_c given in Fig. 6.5, because equation (6.3) is really nothing to do with 'shear stress' at all but is giving additional shear strength by providing quite a different system.

To avoid the muddle which sometimes arises when people talk inaccurately of links being used to meet shear stresses, which we have just seen they certainly do not, and cannot, we shall henceforth in this book refer to shear *stresses* no more, but always discuss shear problems in terms of the imposed *shear force V* arising from ultimate loads, and the *total shear strength* of the beam to resist that shear force. Thus we can add the strength arising from v_c given in Fig. 6.5, to the strength arising from the use of links, and no confusion will arise.

Bent-up bars

Earlier in this chapter we mentioned that an obvious way of preventing diagonal tension cracks forming in the beam would be to have steel bars bent-up at 45° to cross the planes where such cracks would otherwise develop. This is our last resort in the matter, and very effective too; nevertheless such reinforcements involve practical problems in bending and holding the bars in place in the formwork prior to concreting. A certain amount of links are required in any case to tie the whole of the beam together and help steady the longitudinal bars in position, and these are always brought into account first in preference to the use of bent-up bars. CP.110 (1972) now limits the reliance that may be placed on bent-up bars to be no more than the shear contribution provided by the links.

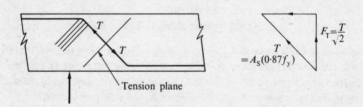

FIG. 6.7. Bent-up bars.

Consider the arrangement of bars bent-up at 45° as in Fig. 6.7. If A_s is the area of the bars, and $0.87 f_y$ is the ultimate stress to be allowed in them, then the total inclined tension in the bars is $A_s (0.87 f_y)$ and the useful vertical component of this force is

$$F_T = \frac{A_s(0.87\,f_y)}{\sqrt{2}} = 0.7\,A_s(0.87\,f_y).$$

Looking back to Fig. 6.1 we remember that any inclined tension force has always to be associated with a complementary inclined compression force, so we have also a vertical compression component of

$$F_C = 0.7\,A_s\,(0.87\,f_y).$$

Thus the introduction of our $45°$ bent-up bars has provided a strength that would in any normal case resist an ultimate shear force applied to the beam of

$$V = F_T + F_C$$

$$= 1.4\,A_s(0.87\,f_y). \tag{6.4}$$

In this way, the bent-up bars have served to assist lift the natural inclined arching action up to the top of the beam.

One great merit bent-up bars have over links is the convenience with which they can be anchored at the top or bottom of the beam by running them horizontally some suitable distance; indeed normally their horizontal continuation is then taken advantage of to provide longitudinal tension strength to resist bending.

Links, it has to be realized, also need good anchorage at the top and bottom of the beam; and this is achieved partly by the development of bond† between the link and the concrete, and partly by passing the links round the main longitudinal bars which, being generally much larger diameter than the links, are able to gain support from the concrete by dowel action. In continuous beams, where the effect of negative bending produces longitudinal tension and cracking in the concrete at the top of the beam, the bond anchorage is likely to be not so good, and for this reason it is particularly desirable the links should be slung from substantial top bars. In this way the proper truss-action of the links is assured. The top bars at the supports of continuous beams are required in any case to provide the tension strength the beam requires to resist bending.

Shear-force diagrams

Before we give examples to illustrate how beams may be designed so they are strong enough to resist specific shear forces, it is necessary to

---

†Bond is referred to further in Chapter 11.

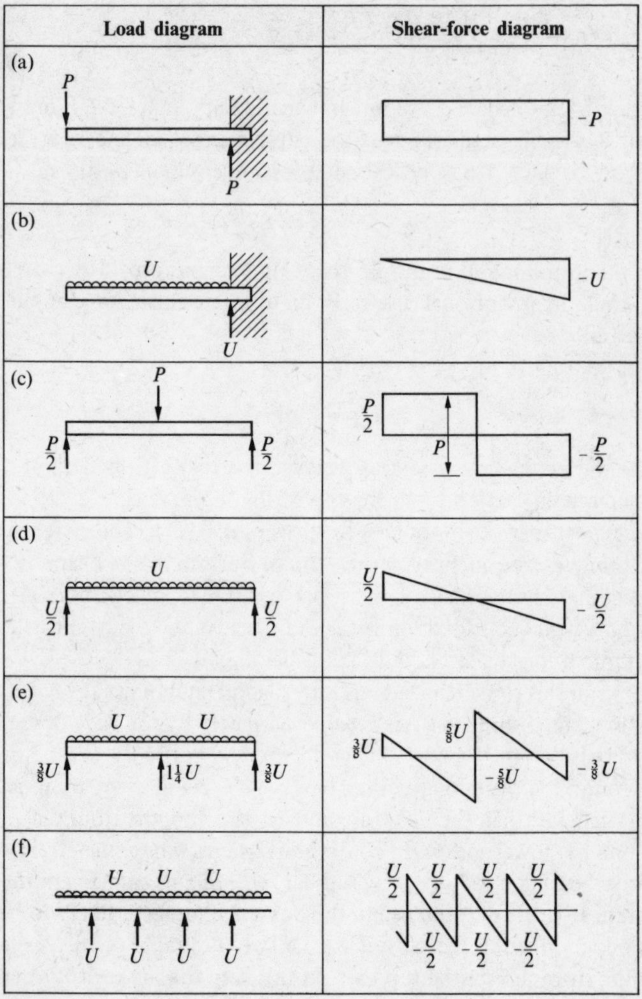

Load diagram	Shear-force diagram

FIG. 6.8. Typical shear-force diagrams.

explain the typical shear-force diagrams given in Fig. 6.8. Note that in considering shear, we are dealing only with *forces*; we are not concerned here with distances or lever arms as we were in Fig. 5.1 with bending moments. In general terms, the shear force at any point on a beam is derived by adding up all the forces which act to the left of the point being considered, counting the upward ones as positive and the downward ones as negative.

Thus for the cantilever carrying an end point-load P (Fig. 6.8(a)), the shear force at all positions along the beam is $-P$.

For the cantilever carrying a total uniform load U, the shear force at the left-hand end is zero, and the shear force at the right-hand end is $-U$. At the mid-point the shear force due to the half-load to the left is $-(U/2)$. Hence the shear force varies in triangular fashion as shown at Fig. 6.8(b).

For the beam with the central point-load P (Fig. 6.8(c)), the two support reactions are both equal to $P/2$. Thus the shear force for the left half of the beam is $P/2$ and for the right half is $-(P/2)$, the change of shear at the centre point-load being equal to P, the amount of the load.

For the beam with the uniform load U (Fig. 6.8(d)), the two support reactions are equal to $U/2$, so the extreme end shears are $U/2$ at the left end and $-(U/2)$ at the right end; and these shears reduce in triangular fashion, becoming zero at midspan, the support reaction acting upwards $(U/2)$ then being exactly matched by half the applied load acting downwards $(-U/2)$.

The following design examples will illustrate how an engineer makes use of (a) longitudinal bars, (b) links, and (c) bent-up bars when designing a beam to achieve a specific shear strength.

Example 1

Consider the beam, 600 mm x 250 mm, shown in Fig. 6.9, simply supported and carrying a characteristic load of 28 tonnes, including its own weight over a span of 5 metres.

Design load $U = 1\frac{1}{2} \times 28 = 42$ tonnes $= 420$ kN.

Maximum shear force $V = \dfrac{420}{2} = 210$ kN.

Maximum bending moment $M = \dfrac{420 \times 5}{8} = 260$ kN m.

Thus $\qquad \dfrac{M}{bd^2} = \dfrac{260 \times 10^6}{250 \times 525^2} = 3\cdot8$ N/mm^2

so from the Chart in Fig. 4.7 we read

$$\dfrac{100\,A_s}{bd} = 1\cdot40,$$

giving $\qquad A_s = \dfrac{1\cdot40 \times 250 \times 525}{100} = 1840$ mm^2.

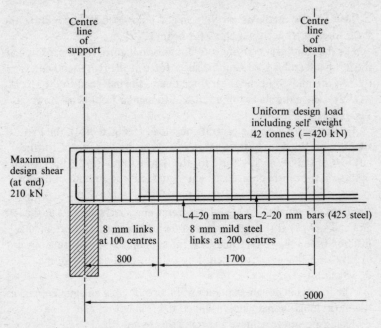

FIG. 6.9. Example 1: Shear reinforcement of a simply-supported beam.

This is provided by 6–20 mm dia bars of 425 steel (area = 1890 mm²).

Note that d has been taken at midspan at 525 mm because 6 bars have to be disposed into 2 layers. At the support, where shear considerations become important, only 4 bars will continue, so there we can take d = 550 mm.

Strictly speaking, to determine v_c for our beam, (the ultimate shear stress taken on the concrete), we should work out the percentage of longitudinal tension reinforcement at each section considered. For simplicity however we shall limit our attention to the case of 4 bars, so that

$$\frac{100\,A_s}{bd} = \frac{100 \times 1260}{250 \times 550} = 0.92,$$

whence from Fig. 6.5 we have v_c = 0·63 N/mm²,
which gives us a shear strength V_c = 0·63 × 250 × 550   = 87 kN.
For the links we shall use 8 mm mild steel, which is convenient to handle, and provides a cross-sectional area

A_{sv} of 101 mm^2; and if we space these at 200 mm centres, they will give a percentage of

$$\frac{101}{250 \times 200} = 0.2,$$

which is a correct minimum.

From equation (6.3) the strength from these mild steel links

will be $\dfrac{A_{sv}(0.87\,f_{yv})\,d}{s_v} = \dfrac{101 \times (0.87 \times 250) \times 550}{200}$ = 62 kN.

giving a total strength of   149 kN.

Now the shear force at the end of the beam is 210 kN, so the 8 mm links at 200 mm centres will be adequate until we reach a point from the centre of the beam equal to $\dfrac{149\ kN}{210\ kN} \times 2500\ mm = 1770\ mm$

(see Fig. 6.8(d)).

Beyond this we need to bunch the links closer together, and if we try 8 mm links at 100 mm centres we have

V_c from concrete shear strength, as before,   = 87 kN.

plus from links $\dfrac{101 \times (0.87 \times 250) \times 550}{100}$ = 124 kN.

giving a total shear strength of   211 kN.

This matches the ultimate shear load very well and is therefore satisfactory.

Example 2

Consider now the beam in Fig. 6.10, the same size as in our previous example, but part of a structural frame which gives it continuity at its ends. Retaining the same 6—20 mm dia bars in the bottom, the load on the beam can now be increased by 50 per cent (by virtue of the continuity). The end shear is therefore increased from 210 kN to 315 kN, and clearly 8 mm dia links would have to be squeezed impracticably close. Accordingly we shall use 10 mm.

As in the previous example we have V_c = 87 kN.

From equation (6.3), 10 mm links at 250 mm centres will provide

$$\frac{157 \times (0{\cdot}87 \times 250) \times 550}{250}$$ = 75 kN.

                    162 kN.

This will take us to $\dfrac{162 \text{ kN}}{315 \text{ kN}} \times 2500 \text{ mm} = 1280 \text{ mm}$ ·

from the centre of the beam.

Beyond this we can double the number of links, spacing them at 125 mm.

Then we get, as before, V_c = 87 kN.

Plus provided by links   = 150 kN.

                    237 kN.

This will take us to $\dfrac{237}{315} \times 2500 \text{ mm} = 1860 \text{ mm}$

from the centre of the beam.

Beyond this point we can conveniently bend up two of the 20 mm longitudinal bars at 45°, and add the strength from these to the value of the links at 125 mm centres which we shall continue through to the column face.

Then we have, as before, V_c = 87 kN.

Plus provided by links   = 150 kN.

From equation (6.4), the effect of 2–20 mm bent-up bars will be
$1{\cdot}4\, A_s(0{\cdot}87\, f_y)$

$$= \frac{1{\cdot}4 \times 628 \times (0{\cdot}87 \times 425)}{10^3}$$ = 325 kN.

                    562 kN.

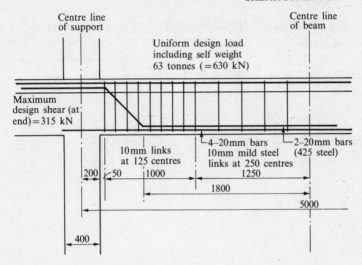

FIG. 6.10. Example 2: Shear reinforcement of a continuous beam.

This is well in excess of the ultimate shear load, and therefore satisfactory.

Note that the balance of shear load that could not be taken on the concrete by inclined tension was 315 − 87 = 228 kN, and we have complied with the requirement of CP.110 (1972) by taking at least half this balance on the links.

Remember also that Fig. 6.1 showed us that shear forces are resisted by diagonal tensions acting in conjunction with diagonal *compressions*. We pointed out at the beginning of the chapter that because concrete is about ten times as strong in compression as in tension, it is normally the tension stresses that are most likely to worry us. Nevertheless when a beam is very heavily reinforced to meet shear forces, it is possible for the compression planes to become overstressed. If we consider a diagonal plane at 45° in just the same way as we did in deriving equation (6.2), it is clear from the same arguments that the *compressive* shear stress is also given by

$$v = \frac{V}{bd}.$$

However, as in practice the compressions are often concentrated in bands only about one third the depth of the beam (see Figs. 6.4 and 6.6), the maximum compressive shear stress in the ultimate condition, using this formula, should likewise be reduced to about a third the value given at equation (2.5). Thus for our Grade 25 concrete we limit the compressive shear stress to a third of $11 \cdot 2$ N/mm^2 equals 3.75 N/mm^2. In the present example we have

$$v = \frac{V}{bd} = \frac{315 \times 10^3}{250 \times 500} = 2 \cdot 5 \text{ N/mm}^2$$

and is therefore quite satisfactory.

7

Columns

The previous chapters were about beams and slabs spanning horizontally between suitable supports. The present chapter is to do with columns, which are the upright members in a structure, and which serve to carry the loads from the various levels down to the foundations.†

It is clear that a concrete column would in some cases be quite satisfactory as a compression member without the inclusion of any reinforcement, since concrete is strong in compression; and in certain cases it might be justifiable to build columns of plain concrete in this way. In practice, however, it is normal to reinforce columns, partly to cater for the effects of bending (discussed later in this chapter), but also for the following reasons, which arise whether the column is subjected to bending or not.

(a) When the columns are long in relation to their diameter, there is a tendency for them to buckle under load. This buckling tends to produce tension on one face, and if reinforcement is provided this tension may be resisted quite safely, and the tendency to buckle be obviated; whereas, if no steel were provided on this tension face the tendency to buckle would not be resisted, and failure would occur.

(b) Where there is the slightest risk of unequal settlement, or unequal expansion and contraction, it is always possible that under certain conditions a few columns in a large complicated structure might actually be under tension when the surrounding floors were unloaded above and heavily loaded below. This would, of course, cause cracking across the columns, and can be entirely prevented by vertical reinforcing bars.

(c) Where the size of the column to carry the necessary load becomes excessive, it can be materially reduced by adding a sufficient amount of steel reinforcement, since, as indicated in the early chapters, the compressive design strength of high-yield steel is about twenty-five

---

†There is a companion book written by John Faber and Brian Johnson entitled *Foundation Design Simply Explained* to which the reader is referred.
Publisher: Oxford University Press.

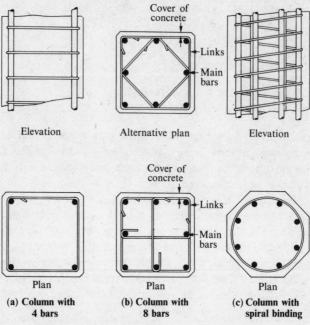

FIG. 7.1. Typical arrangements of column reinforcement.

times that of concrete (depending of course on the exact type of steel and grade of concrete), and therefore for each square mm of reinforcement we have the same load-carrying capacity as there would be in 25 mm^2 of concrete.

For these reasons columns are invariably reinforced with vertical bars. CP.110 (1972) requires generally not less than 1 per cent, nor more than 6 per cent of the gross cross-sectional area of the concrete. Fig. 7.1 shows some typical arrangements. Diagram (a) with four bars is the most common form of reinforced column up to about 300 mm square; beyond that eight bars are often used, up to about 500 mm square. The vertical bars generally vary from about 16 mm diameter up to about 32 mm diameter.

It will be noticed that the bars are placed near the edges of the column rather than towards the centre. This is partly because in these positions they act best in preventing the column as a whole from buckling. But the main reason is that here they are most effective in resisting tension from bending; and this is discussed in detail later in this chapter. Where bending is a major consideration about one axis rather than the other, it is quite usual to make the column rectangular in plan so as to enhance its strength about that axis.

It should be noticed too that because the bars are placed near the edges of the column, and are in general very long in comparison to their diameter, they would be extremely liable themselves to buckle outwards, bursting the small cover of concrete away from the column, and so destroy it. To prevent this the vertical bars are held in by means of transverse links, clearly shown in Fig. 7.1. These links are generally 6 mm or 8 mm diameter, spaced at about 200 mm intervals up the column.

Where four vertical bars only are used, square links are all that are necessary; but where eight bars are used square links would not be effective in preventing the bars half-way along the face from bursting outwards, since considerable movement could take place before the links became effective. Therefore additional links have to be used, as indicated. For circular or octagonal columns, the transverse reinforcement is normally in the form of a spiral binding; this considerably strengthens the concrete by acting as a corset, preventing it bulging outwards.

Centrally loaded columns

Consider the symmetrical reinforced concrete column shown in Fig. 7.2. The net cross-sectional area of the concrete is A_c, and the total cross-sectional area of the reinforcing steel is A_{sc}. If a load N† is applied precisely at the centre of the column, and the column has been constructed truly so as to coincide exactly with the line of the load, there is no difficulty in writing the equation for N in terms of the stresses in the concrete and the steel. The design strength for the concrete is $0.45 f_{cu}$ (see equation (2.5)); and the design strength for the steel is $0.74 f_y$ (see equation (2.7)), so we have

$$N = \text{concrete force} + \text{steel force}$$
$$= 0.45 f_{cu} A_c + 0.74 f_y A_{sc}.$$

However, in practice, owing to minor errors of construction, it is inevitable that the line of application of the load will fail to coincide exactly with the axis of the column, so that the stresses across the

---

†N is the symbol used in CP.110 (1972) for the ultimate central load on a column.
For beams, the authors have preferred to use U for uniform loads, and P for point loads. So long as the differences are clear, no confusion should result.

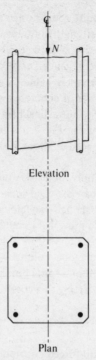

Elevation

Plan

FIG. 7.2. Column subject to direct compression.

section will not be quite uniform, those at one edge being a little more than average and those at the opposite edge correspondingly less.

For this reason we reduce by 10 per cent the stresses that we design to, so we arrive at a realistic value for the design-load on a supposedly centrally loaded column as

$$N = 0 \cdot 40 f_{cu} A_c + 0 \cdot 67 f_y A_{sc}. \qquad (7.1)$$

This formula is as given in CP.110 (1972) and allows for a lack of alignment of the load (what we term the *eccentricity*) of one-twentieth the overall width of the column.

In most real situations, the load on a column arrives through the various reinforced concrete members framing into the column, as for example the beams and slabs of a floor or roof in a multi-storey building. If the column is situated internally so as to support a symmetrical configuration of beams, the condition of central loading is achieved provided the beams on opposite sides of the column are of equal span and carrying equal loads. This is not a very likely situation, partly because in real buildings it is difficult to arrange for the beam spans to be identical, and also because it is unlikely the imposed loads on the floor would always be in balance about the column. Nevertheless, provided the beam spans on opposite sides of the column do not differ by more than 15 per cent, and provided the imposed loads on the beams are uniformly distributed (as opposed to being point-loads), we can make an approximate assessment that the column will be satisfactory so long as we reduce the stresses in the concrete and steel a further 10 per cent below the values given in equation (7.1).

Thus we have

$$N = 0 \cdot 35 f_{cu} A_c + 0 \cdot 60 f_y A_{sc}. \qquad (7.2)$$

This is a good practical formula for use in designing most internal columns in multi-storey framed buildings. A simple example will demonstrate its use.

Let us take the case of a 300 mm square column reinforced with four 25 mm dia. bars. The gross cross-sectional area of the column is 90 000 mm^2, and the area of steel is 1960 mm^2; so the percentage of steel is about 2·2 per cent, well above the minimum requirement of 1·0 per cent. We will design for our Grade 25 concrete, and 425 N/mm^2 cold-worked high-yield steel.

We have $A_{sc} = 1960$ mm^2,

so $A_c = 90\ 000 - 1960 = 88\ 040$ mm^2

therefore $N = 0 \cdot 35 f_{cu} A_c + 0 \cdot 60 f_y A_{sc}$

$\qquad = (0 \cdot 35 \times 25 \times 88\ 040) + (0 \cdot 60 \times 425 \times 1960)$

$\qquad = 770\ 000 + 500\ 000$

$\qquad = 1\ 270\ 000$ N $= 1270$ kN or 127 tonnes.

Note the considerable proportion of the design load the steel carries with only 2·2 per cent. The permitted limits for steel are 1·0 and 6·0 per cent, so it is clear we can obtain a considerable range of strengths just by varying the amount of steel we provide in a constant over-all size of column. This is very convenient, because standardization of column sizes leads to repetition of column formwork at the various floor heights up a building; it also makes for repetition of beam formwork (the beams not getting longer as construction proceeds to the higher floors). Such standardization also facilitates repetition of partition units and other similar builders items.

Equations (7.1) and (7.2) apply only to columns which are not long enough in relation to their width to be subject to any appreciable risk of buckling. The same is true of the further equations derived later in this chapter. It is not easy to define exactly the proportions at which columns become so slender that they constitute a buckling risk; sufficient for the present book, however, it may be said that, as a general rule, when the length does not exceed sixteen times the least lateral dimension, there will be little risk of buckling, provided the column has the minimum percentage of vertical steel of 1·0 per cent, and the column is properly held in place at both the top and the bottom. This ratio of 16 is generally sufficient to cover most cases met in normal building construction.

Columns subjected to bending

In practice it is very rare to achieve true central (*concentric*) loading, except in the case of internal columns where the arrangements of columns and beams and loading are all symmetrical. In the case of columns at the outside edge of a building concentric loading is virtually impossible.

As will be seen from Fig. 7.3, a reinforced concrete beam, when it

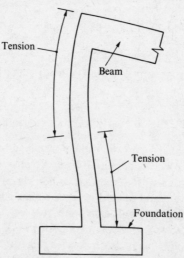

FIG. 7.3. Single-storey column bent by deflection of beam (exaggerated).

deflects, necessarily bends, and when it is constructed continuously (i.e. is monolithic) with the column, the reinforced concrete joint at the end will be stiff, and therefore deflection of the beam will necessarily produce bending in the column. The diagram in Fig. 7.3 is, of course, grossly exaggerated to make the point clear, and in practice the bending of both column and beam under loading is so slight as not to be visible to the naked eye though both are susceptible to measurement by delicate instruments. It is, however, quite a mistake to assume that the bending does not occur merely because it cannot be seen; in the same way that it must be remembered that shortening of a column due to the direct stress is also so slight that it cannot be seen with the naked eye, yet no one would venture to suggest in consequence that the column is not subjected to direct stress.

Fig. 7.3 shows that the column, instead of being compressed everywhere, may be stressed in tension on one face for the upper portion and on the opposite face for the lower portion, and it is clear that when part of the column is in tension the total compression which it can safely carry is very considerably reduced. Nevertheless, the column is made much safer if vertical reinforcing bars are put in capable of resisting all tension of this kind, since if a similar column were stressed in the same way without reinforcement it would simply crack across the tension plane.

If the column is very stiff compared to the beam, it will offer great restraint to the end of the beam. In the extreme case, when the stiffness of the column is infinite compared with that of the beam, the case becomes that of a beam *encastré* (that is to say, rigidly held in direction at the ends), for which the end bending moment resulting from a uniformly distributed load is $UL/12$. In the case of such a very stiff top-storey column supporting a roof beam, so that the column does not extend above (see Fig. 7.4), it is clear the whole of this bending moment has to be carried by the column section under the beam, so that the moment in the column is also $UL/12$. If, however, the column is at any of the storey-heights below, extending above and below the beam and of equal length above and below, then obviously this bending moment of $UL/12$ will be shared equally between the top and bottom sections, so that the bending moment in the column will be $UL/24$. Anyone wishing to avoid complication in the design of outside columns and willing to design for ample moments can therefore apply the above moments without further complication. They will always be on the right side, but they will in most circumstances be considerably greater than necessary.

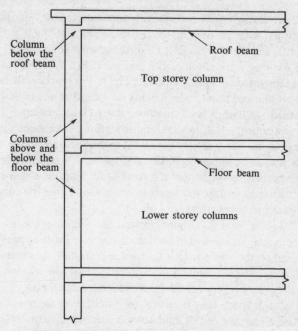

FIG. 7.4. External column of a multi-storey frame.

Where the column is flexible compared to the beam, it will offer correspondingly less restraint, and the bending moment in the columns will then lie between zero and the figures given above. In determining more accurately the moments in such intermediate cases, a mathematical consideration of the problem shows that the determining factor is the relative stiffnesses of the incoming beam and the column sections immediately above and below, in accordance with the expressions given in Table 7.1.

The symbols used have the following meanings.

M_e is the bending moment which would occur at the end of the beam where it joins the column, if, with its particular system of loading, it were *encastré* or rigidly held in direction. In the case of uniform loading, this end-moment may be taken as $UL/12$: with a single point-load at midspan it may be taken as $PL/8$.

K_u is the stiffness of the column *above* the floor under consideration. K_l is the stiffness of the column *below* the beam under consideration. K_b is the stiffness of the beam. In each case the stiffness is arrived at by the moment of inertia of the member divided by its length (that is to say I/L).

TABLE 7.1
Moments in external columns (and similarly loaded columns).

Moment at foot of upper column	$M_e \dfrac{K_u}{K_l + K_u + 0 \cdot 5\, K_b}$
Moment at head of lower column	$M_e \dfrac{K_l}{K_l + K_u + 0 \cdot 5\, K_b}$

The moment of inertia of a simple rectangular (or square) section is given by the formula

$$I = \frac{bh^3}{12},$$

where b is the breadth, and h is the depth. For T-beam sections the calculation is not as simple, and the reader is referred for convenience to the worked-out values given in Table 5.2. In either case the effect of the steel reinforcement in the small percentages we are concerned with may safely be ignored.

Examples of the use of Table 7.1 in calculating the moments in an actual column for a multi-storey building are given towards the end of this chapter. But before studying these, it will be appropriate first to derive the necessary equations for assessing the suitability of any column to cope with such bending moments in combination with direct compressive loading.

Bending and direct compression combined

If a column of breadth b, and depth h, is subjected to a bending moment M at the same time as it carries a direct load N, we can for analytical purposes regard this as being identical to the load N being applied at an eccentricity from the centre of the column equal to $e = M/N$. This is illustrated in Fig. 7.5.

Fig. 7.6 shows the reduced effective compressive stress block which is available to resist the load N when it acts at the eccentricity e. If we ignore the effects of the reinforcing steel, this stress block has to be equally disposed about the line of action of N; and if N is a distance $(h/2 - e)$ from the face of the column, the depth of the effective stress block will be $2(h/2 - e)$, or more simply $(h - 2e)$. Thus if the column width in the other direction is b, equation (7.1) becomes

$$N = 0 \cdot 40\, f_{cu} b(h - 2e). \tag{7.3}$$

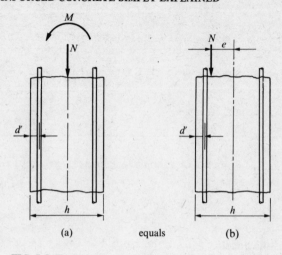

FIG. 7.5. Typical column: load, moment and eccentricity.

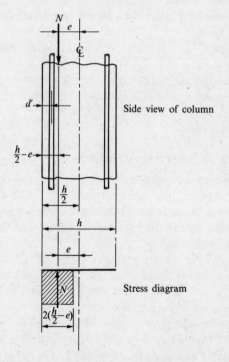

FIG. 7.6. Combined bending and direct-compression in a column: simple case.

This rough and ready formula completely ignores the contribution of the reinforcing steel.

When e coincides with the face of the column, the extent of the stress block is nil, so the stress would be infinity and hence absurd. Accordingly a strict limit has to be put on the use of equation (7.3) to cases where e does not come outside the line of the reinforcing steel, i.e. e must not be more than $(h/2 - d')$, where d' is the distance in to the steel on the compression side. Even so the stress block will then be only $2d'$, which is little enough, and certainly only satisfactory if the minimum provisions of vertical and transverse reinforcements are complied with. Equation (7.3) has been shown here primarily for interest's sake, since it introduces the reader to the idea of how bending in a column can be regarded as causing an eccentricity of the load, and how the result of this is to reduce the effective proportion of the section actually carrying the direct stress.

Let us now consider in more detail the general case shown in Fig. 7.7 of a column $b \times h$ with a total area of steel A_{sc} symmetrically disposed. If the load N is applied centrally ($M = 0$ and thus $e = 0$), the concrete will be stressed uniformly and the stress in each of the bars will be equal. However when we apply a moment to the column as shown, this has the same effect as if N were moved to some point left of centre, so that the $s1$ bars take more than an equal share of compressive stress and the $s2$ bars take less. If the eccentricity is more than a certain amount, the stress in the $s2$ bars will change from compression to tension, so that the section behaves very much like a beam as we saw in Chapter 4, with the compression in the concrete not reaching the right-hand side of the section, and a neutral axis develops in some position as shown in the Figure.

When we were considering columns loaded roughly centrally (equations (7.1) and (7.2)), and when the eccentricity was small enough so that tension in the $s2$ bars could not occur (equation (7.3)), we were able to calculate the stresses simply by use of a single formula; but now if we seek to analyse the case where there is no limit on the amount of eccentricity, so that the effects of the reinforcements clearly have to be taken into account, we shall find that we are faced with a situation which can only be solved by a process of trial and error. This is because the position of the neutral axis will vary for a given combination of N and e depending on the area of steel we provide. Very wisely CP.110 (1972) requires d_c (the distance to the neutral axis) never to be less than $2d'$, much like the restriction we saw in the use of equation (7.3).

Let us now state the facts of Fig. 7.7 algebraically.

First we resolve all forces acting vertically as follows.

The concrete stress acts on the area bd_c
and can provide a force of

$$0.4 f_{cu}bd_c$$

The $s1$ bars will always be in compression.

Stressed at $0.72 f_y$† on the area $\dfrac{A_{sc}}{2}$ they

can provide a force of

$$0.72 f_y \frac{A_{sc}}{2}.$$

The $s2$ bars may be in compression stressed
up to $+0.72 f_y$ (if they come within the
shaded area of Fig. 7.7), or may be in
tension stressed up to $-0.87 f_y$ (if they
come outside the shaded area). Denoting
whichever of these is applicable as
f_{s2}, they provide a force (compressive
or tensile) up to

$$f_{s2} \frac{A_{sc}}{2}.$$

Adding these 3 terms together we get

$$N = 0.4 f_{cu}bd_c + 0.72 f_y \frac{A_{sc}}{2} + f_{s2} \frac{A_{sc}}{2}. \qquad (7.4)$$

Secondly we take moments of these 3 forces about the centre-line of
the section which gives us

$$M = 0.4 f_{cu}bd_c \times \left(\frac{h}{2} - \frac{d_c}{2}\right) + 0.72 f_y \frac{A_{sc}}{2} \times \left(\frac{h}{2} - d'\right) + f_{s2} \frac{A_{sc}}{2} \times \left(\frac{h}{2} - d'\right)$$

or more simply

$$M = 0.2 f_{cu}bd_c(h - d_c) + 0.72 f_y \frac{A_{sc}}{2} \left(\frac{h}{2} - d'\right) + f_{s2} \frac{A_{sc}}{2} \times \left(\frac{h}{2} - d'\right). \qquad (7.5)$$

---

† Actually for the steel stress of 425 N/mm² we are using throughout this book,
$0.74 f_y$ would be quite suitable, as given in eqn (2.7). However CP.110 (1972),
uses $0.72 f_y$ in this context so as to cover for the extreme and rare use of
460 N/mm² steel in columns; and the same figure has been adopted here to avoid
confusion.

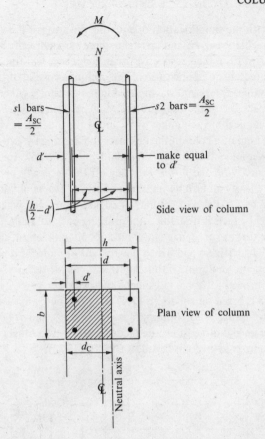

FIG. 7.7. Combined bending and direct-compression in a column: general case.

These equations, (7.4) and (7.5), are the basic general equations for the design of columns subjected to combined bending and direct compression whatever the eccentricity. The only restriction on their use is that d_c must never be less than $2d'$. In CP.110 (1972) the same equations are given except that the matter has been made more complicated there by making provision for the s1 bars being different in size and position from the s2 bars. This seems an unnecessary refinement, since (as portrayed by Fig. 7.3) tension can sometimes occur on either face of a column, and errors are far more likely to creep into design and construction work if we start making things more complicated and fussy than necessary.

Even with the simplification of equating the areas of the $s1$ and $s2$ reinforcements to $A_{sc}/2$, and arranging these symmetrically an equal distance d' in from their respective faces, the reader will find the equations tedious to use because of the options available of trying different values for d_c and A_{sc}. Indeed in the examples which follow we shall make use of the standard design chart given at Fig. 7.8 which takes a great deal of the donkeywork out of the process. It will be seen that the design of a column then reduces to choosing by experience a suitable over-all section size ($b \times h$), and for the known values of N and M (and hence N/bh and M/bh^2) reading off the necessary percentage of steel A_{sc}. Clearly an infinite number of answers can be obtained, depending on what size of column a designer wants, and what percentage of steel he feels is desirable. And, turning back to Table 7.1, we remember that as we change our choice of column section, so shall we be altering its stiffness and hence the amount of moment it carries. The task of designing columns with bending is therefore seen to be not without the need for some judgement and patience.

Fig. 7.8 is based on equations (7.4) and (7.5) but is for use only with columns where $f_{cu} = 25$ N/mm^2, $f_y = 425$ N/mm^2, and $d/h = 0.90$. For other stresses and dimensions the reader is referred to others of the family of charts as given in Part 2 of CP.110 (1972).

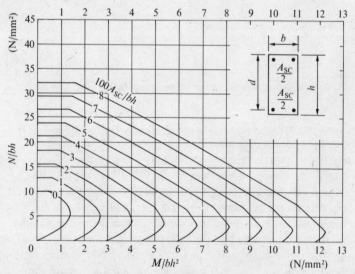

FIG. 7.8. Standard design-chart for a column subject to combined bending and direct-compression. (Grade 25 concrete and 425 steel).

Example

Let us now consider the case of the external column of a multi-storey building as shown at Fig. 7.9, where we shall take the column lengths as being 3·5 metres. We start at roof level, where the incoming beam has a span of 7·5 metres and carries a uniform characteristic load of 25 tonnes including its own weight. The beam design load is therefore

$$U = 1\tfrac{1}{2} \times 25 = 37\cdot5 \text{ tonnes} = 375 \text{ kN},$$

and the approximate midspan bending moment is

$$M = \frac{UL}{11} = \frac{375 \times 7\cdot5}{11} = 256 \text{ kN m}.$$

From Table 5.2 we select a T-beam size 500 mm × 200 mm as being suitable; and taking its moment of inertia as $3\cdot91 \times 10^9$, we have its stiffness

$$K_b = \frac{I}{L} = \frac{3\cdot91 \times 10^9}{7\cdot5 \times 10^3} = 0\cdot52 \times 10^6.$$

For the top storey we will first try a typical column 400 mm × 400 mm for which

$$I = \frac{400 \times 400^3}{12} = 2\cdot13 \times 10^9 \text{ (as in Table 7.2)},$$

so the column stiffness is

$$K_1 = \frac{I}{L} = \frac{2\cdot13 \times 10^9}{3\cdot5 \times 10^3} = 0\cdot61 \times 10^6.$$

Now the encastré moment at the end of our beam is

$$M_e = \frac{UL}{12} = \frac{375 \times 7\cdot5}{12} = 234 \text{ kN m},$$

whence from Table 7.1 we have the moment at the head of our roof column

$$M = M_e \frac{K_1}{K_1 + K_u + 0\cdot5\,K_b}$$

$$= 234 \frac{0\cdot61}{0\cdot61 + 0 + (0\cdot5 \times 0\cdot52)} \cdot \frac{10^6}{10^6}$$

$$= 164 \text{ kN m}.$$

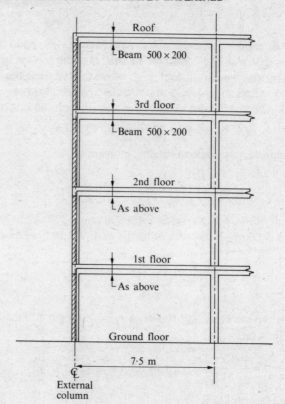

FIG. 7.9. Example of external column in multi-storey building.

Remembering we use N as the symbol for design load in columns, we have N equal to $\frac{1}{2}$ x 375 = 188 approximately. Now we need to work out N/bh and M/bh^2, so we can read off from Fig. 7.8 the percentage of vertical steel (A_{sc}) required in the column. This is made a little easier for us by having bh and bh^2 available from Table 7.2, so we have

$$\frac{N}{bh} = \frac{188 \times 10^3}{160 \times 10^3} = 1\cdot 18 \text{ N/mm}^2$$

$$\frac{M}{bh^2} = \frac{164 \times 10^6}{64 \times 10^6} = 2\cdot 56 \text{ N/mm}^2.$$

From Fig. 7.8 we read off that the percentage of steel required is only 1·3 per cent, so we shall try the next smaller size of column which is 350 mm x 350 mm.

TABLE 7.2
Properties of typical square columns

Column size b × h (mm × mm)	Gross section area bh (mm² × 10³)	bh² (mm³ × 10⁶)	Moment of inertia of gross section (mm⁴ × 10⁹)	Vertical reinforcement					
				Total no of bars and dia: and percentage			No of bars effective in bending: and percentage		
				Light (mm)	Medium (mm)	Fairly heavy (mm)	Light (mm)	Medium (mm)	Fairly heavy (mm)
200 × 200	40·0	8·0	0·13	4–12 1·12%	4–16 2·00%	4–20 3·15%	As for total no of bars as in previous columns (except where noted below)		
250 × 250	62·5	15·6	0·32	4–16 1·28%	4–20 2·02%	4–25 3·15%			
300 × 300	90	27	0·67	4–20 1·40%	4–25 2·18%	4–32 3·57%			
350 × 350	123	43	1·25	4–20 1·02%	8–20 2·05%	8–25 3·18%		6–20 1·54%	6–25 2·38%
400 × 400	160	64	2·13	4–25 1·23%	4–32 2·01%	8–25 2·45%			6–25 1·84%
450 × 450	203	91	3·42	8–20 1·23%	8–25 1·92%	8–32 3·15%	6–20 0·93%	6–25 1·44%	6–32 2·38%
500 × 500	250	125	5·20	8–20 1·00%	8–32 2·57%		6–20 0·74%	6–32 1·92%	

Proceeding as before, we have

$$K_1 = \frac{1 \cdot 25 \times 10^9}{3 \cdot 5 \times 10^3} = 0 \cdot 36 \times 10^6,$$

$$M = 234 \frac{0 \cdot 36}{0 \cdot 36 + 0 + (0 \cdot 5 \times 0 \cdot 52)} \cdot \frac{10^6}{10^6}$$

$$= 136 \text{ kN m.}$$

$$\frac{N}{bh} = \frac{188 \times 10^3}{123 \times 10^3} = 1 \cdot 51 \text{ N/mm}^2$$

$$\frac{M}{bh^2} = \frac{136 \times 10^6}{43 \times 10^6} = 3 \cdot 16 \text{ N/mm}^2.$$

From Fig. 7.8 we find that the steel required is still only 1·8 per cent. This is because the smaller size of column, being less stiff, has offered less restraint to the end of the beam, and the bending moment taken by the column is 136 kN m instead of 162 kN m, a reduction of about 15 per cent.

Let us now try a column size of 300 mm x 300 mm. Proceeding as before, we have

$$K_1 = \frac{0 \cdot 67 \times 10^9}{3 \cdot 5 \times 10^3} = 0 \cdot 19 \times 10^6$$

$$M = 234 \frac{0 \cdot 19}{0 \cdot 19 + 0 + (0 \cdot 5 \times 0 \cdot 52)} \frac{10^6}{10^6}$$

$$= 99 \text{ kN m.}$$

$$\frac{N}{bh} = \frac{188 \times 10^3}{90 \times 10^3} = 2 \cdot 09 \text{ N/mm}^2$$

$$\frac{M}{bh^2} = \frac{99 \times 10^6}{27 \times 10^6} = 3 \cdot 64 \text{ N/mm}^2$$

giving an area of steel required now of 2·0 per cent.

If the reader were to try the same exercise with a column size of 250 mm x 250 mm he would find that the reinforcement required would be rather heavy, and for other reasons which will emerge shortly, the column of 300 mm x 300 mm will be chosen, reinforced with 4—25 mm vertical bars as read from the sixth column in Table 7.2. The reason for having worked our way studiously through the calculations

of the larger column sizes was to demonstrate the common experience
that in designing *external columns* the effect of going for a larger
section often does little to increase the margin of safety, though of
course the cost increases roughly in proportion to the cross-section, so a
column of 400 mm x 400 mm would cost roughly 1·8 times more than
our chosen column of 300 mm x 300 mm.

Now let us consider the column supporting the topmost floor of our
building, the third floor shown in Fig. 7.9. Earlier in the chapter we
described how a top-storey length of an external column is subjected to
more bending than are the lengths at the levels below; this is because in
the latter cases the bending moment from the beam is shared between
the length of column above and the length of column below. Now in
our real example, we can see the striking effect this has.

It can be taken, for the purpose of our example, that the incoming
beam supporting the third floor is a similar T-beam 500 mm x 200 mm,
7.5 metre span, carrying 25 tonnes characteristic load, i.e. 375 kN
design load, all as before for the roof. In addition the column will have
to carry one storey height of external wall with its supporting beams,
say 8 tonnes characteristic load, i.e. 1½ x 8 = 12 tonnes = 120 kN
further design load. Thus we have

N from beams at roof plus third floor	375 kN
from external wall and beams	120 kN
	495 kN

Note that the external wall and its beams are in balance about the
column, and so produce no additional bending.

We shall try the effect of continuing the same size of column as we
had at roof level, i.e. 300 mm x 300 mm. The moment in the column at
third floor level is therefore

$$M = 234 \frac{0 \cdot 19}{0 \cdot 19 + 0 \cdot 19 + (0 \cdot 5 \times 0 \cdot 52)} \cdot \frac{10^6}{10^6}$$

$$= 70 \text{ kN m.}$$

Thus we have

$$\frac{N}{bh} = \frac{495 \times 10^3}{90 \times 10^3} = 5 \cdot 5 \text{ N/mm}^2$$

$$\frac{M}{bh^2} = \frac{70 \times 10^6}{27 \times 10^6} = 2 \cdot 6 \text{ N/mm}^2,$$

giving an area of steel required of 1·0 per cent. This is half what was required at roof level; the reduction arises from the moment in the column reducing from 99 kN m to 70 kN m.

Now let us try the same column size below second floor level. The beam and wall-loads at this level can be assumed the same as at third floor. Thus we have

N from beams at roof, third and second floor   563 kN

from external wall and beams   240 kN
           —————
           803 kN

$$M \text{ (as before)} = 70 \text{ kN m},$$

$$\frac{N}{bh} = \frac{803 \times 10^3}{90 \times 10^3} = 8 \cdot 9 \text{ N/mm}^2,$$

$$\frac{M}{bh^2} = \text{(as before)} = 2 \cdot 6 \text{ N/mm}^2.$$

Percentage of steel required is 1.7 per cent.

Similarly below first-floor level

$$N = 1110 \text{ kN},$$

$$\frac{N}{bh} = 12 \cdot 3 \text{ N/mm}^2,$$

$$\frac{M}{bh^2} = 2 \cdot 6 \text{ N/mm}^2.$$

Percentage of steel required is 2·5 per cent.

From this we see, amongst other things, that the size of column necessary below the roof is the same as required below first-floor level, notwithstanding the fact that the latter carries six times the load. This is typical experience in the case of external columns, where the bending moment is the greatest at the topmost storey; and it frequently is found to be correct and economical to make the external columns the same size down through many storeys. In the present case, four storeys completely justify this treatment, and had the building been one of greater height, the same procedure would have been justified in one more storey down. This repetition of course makes for speed and economy, enabling the same formwork to be used for the columns and beams without modification; and brickwork and window details, having been worked out once, can be repeated at all levels making for further simplicity and savings of cost.

8

Prestressed concrete

Introduction

We have seen earlier how, with ordinary reinforced concrete, the steel
only starts to work effectively in tension after the concrete around it
has cracked. For the majority of structural applications this is perfectly
satisfactory; and this is the way reinforced concrete is used in most
normal situations.

There are however some circumstances where the use of ordinary
reinforced concrete can lead to certain difficulties. For example, when
slabs and beams are used to span considerable distances, the self-weight
may become appreciably greater than the imposed load to be carried. In
our simple beam described in Chapter 4, half the concrete was in the
tensile zone and thus contributing nothing to the bending strength, yet
it represented half the weight; and as spans get larger and larger, so that
more and more weight arises in the tensile zone, eventually we reach a
stage where the self-weight of the construction becomes so great that
the beam gets too cumbersome even to carry itself.

Another problem that arises with ordinary reinforced concrete is in
members which are repeatedly subjected to shock loads or intense
vibration. After a while, with the tension cracks repeatedly opening and
closing, they eventually fail to fit together properly because of gradual
dislodgement of the fine particles in the crack surfaces. Once this starts,
the rate of wear on the crack surfaces increases, and serious
disintegration results.

Such deficiencies of ordinary reinforced concrete can be overcome
by the use of prestressed concrete. The underlying idea behind
prestressed concrete is to use the steel and concrete in members subject
to bending in such a way that no tensile stresses develop in the
concrete, and hence no cracks. The technique depends on stretching the
steel to very high stresses, and then releasing it so that it acts against the
concrete member to produce a longitudinal compression. The high
stresses in the steel achieve great economy in the quantity of steel
required, though, of course, the steel has to be of a much higher

quality, and, as we shall see, is subject to considerable expenditure on labour and equipment in producing the prestress conditions with the requisite degree of accuracy and control.

When steel is embedded unstressed in an ordinary reinforced-concrete member, care has to be taken in design to ensure that when the member is later put into service the steel is not stressed to a degree that would lead to cracks of objectionable width forming in the concrete. In prestressed concrete, on the other hand, the steel is mechanically stretched and stressed as highly as possible (limited only by the properties of the steel itself), and it is the amount of this stress and accompanying stretch which ensures that the concrete is kept in a state of permanent compression.

Fig. 8.1 illustrates simply how a rectangular beam can be prestressed so that it will be capable of resisting bending without the concrete ever coming into tension. In Fig. 8.1(a) the beam is pre-compressed along its length by jacks acting centrally against both ends so as to produce a

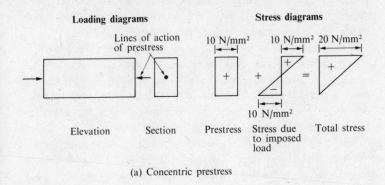

(a) Concentric prestress

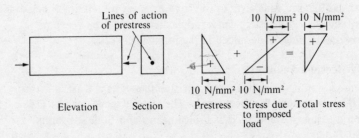

(b) Eccentric prestress

FIG. 8.1. The principle of prestressing.

uniform compressive stress of 10 N/mm^2. If then a bending moment is applied to the beam, such as would produce a compressive stress or 10 N/mm^2 at the top and a tensile stress of 10 N/mm^2 at the bottom, it is clear the resulting stresses in the beam would be 10 + 10 = 20 N/mm^2 at the top and 10 − 10 = 0 at the bottom. We thus have a section which is nowhere in tension; but this absence of tension has been achieved only at the price of doubling the compressive stress at the top.

To overcome this objection, we will now repeat the exercise (as shown in Fig 8.1(b)), but this time lower the point of application of the thrust from the centre to a position one-third up from the bottom. Here, with only half the amount of thrust, we achieve a triangular prestress distribution across the beam section which varies from zero at the top to 10 N/mm^2 at the bottom. When now we apply the same bending moment to the beam as before, the resulting stresses become 0 + 10 = 10 N/mm^2 at the top, and 10 − 10 = 0 at the bottom. Thus, by means of a suitably chosen amount and position of prestress, we have eliminated all tension in the concrete without anywhere causing any increase in the maximum compressive stress.

For normal practical applications, as for example bridge beams, floor slabs, or railway sleepers, it is impractical to apply the end forces by permanent external jacks, and use is made instead of high-tensile steel wires stretched through the length of the members and anchored at both ends. These wires, often 7 or 19 wound cable-fashion, are of characteristic tensile strength generally of the order of 1500 − 1800 N/mm^2, that is roughly four times the strength of normal high-tensile steel bars. The force in the wires creates the required compression in the concrete. By arranging the line of action of one or more of the wires at any position between the centre or the third-point of the unit, we can achieve whichever of the circumstances we require as illustrated by Fig. 8.1. Prestressing can be applied either by *pre-tensioning* or *post-tensioning*, as described in the following.

Pre-tensioned concrete

Pre-tensioning is applied to relatively simple units precast in factories laid out specially for the purpose. The method lends itself to mass production of items such as bridge beams and floor-slab units (Fig. 8.2).

A number of high-tensile steel wires, often 5 mm dia, are stretched between two ground anchorages about 100 or 150 m apart using a

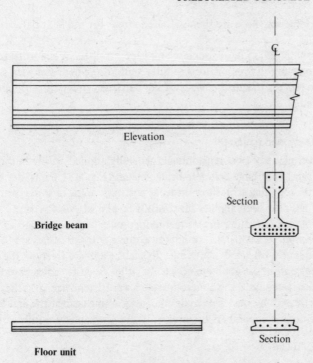

Elevation

Bridge beam

Section

Section

Floor unit

FIG. 8.2. Pre-tensioned units.

hydraulic jack at one end. If the wires are stretched to a stress of about 1200 N/mm^2 (a usual value), then the amount of stretch for every 100 m is about 600 mm. The stress can be determined by the pressure in the ram of the hydraulic jack, which is indicated on a pressure gauge.

Timber or metal moulds are then built along the line of the wires into which the concrete is now poured. The moulds are spaced a short distance apart from one another, so that after the concrete has reached a sufficient strength, the wires can be cut between the units. However the wires have to be kept stretched until the concrete has set sufficiently not only to resist the very considerable compressive stress to which it will be subjected when the wires are cut, but also until the *bond* between the wires and the concrete has developed to such an extent that, when the wires at the end of each unit are cut, the stretched wires do not slip inside the concrete. When eventually the wires are released, the units are carefully moved so that the hardening and curing process can continue under closely controlled conditions;

the prestressing beds are then prepared afresh for the next batch of units.

In the production of simple plank-type members, it is normal to form one continuous casting the full length of the bed, and then saw this to lengths to suit the needs of the particular order currently being fulfilled.

Post-tensioned concrete

The technique of post-tensioning is generally applied on site in the construction of large concrete bridge-beams (Fig. 8.3), dams, dry docks, or water-retaining structures. Indeed, post-tensioning in some form or other has now been applied successfully to almost every type of concrete construction. In this technique, holes or ducts are left when the concrete is cast, and through these ducts wires or cables are subsequently threaded; sometimes the cables are laid freely in the duct-formers before the concrete is cast. In either case the cables have to be left unstressed until the concrete attains a sufficient strength; they are then stressed, by means of hydraulic jacks acting against the end faces of the concrete members themselves. When the cables are fully stressed,

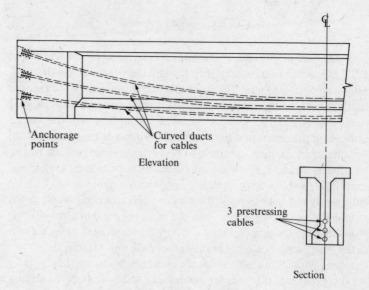

FIG. 8.3. Post-tensioned beam.

they are anchored by wedging devices against the ends of the concrete as described later, and grout is injected under pressure to fill the ducts. The wedges, together with the grout, ensure that the prestressing force is properly 'locked' into the member; the grout also protects the cables against corrosion.

An important development of post-tensioning is the threading of the cables through a number of precast units which are laid out in line and then stressed together. This closely resembles the analogy of carrying a dozen books side by side from one bookcase to another; if you squeeze-in the ends (prestress the twelve books) they act like a single unit, resisting the bending and shears arising from their own weight. Thus a long member like a bridge of 100 m span can be made in segments of about 5 m length, each segment being cast with ducts registering for later receiving the cables which will pass right through from one end of the bridge to the other. The segments are assembled on staging, the cables then drawn through and tensioned and anchored, and the ducts finally grouted all as previously described. The compressive stresses created by the prestressing force produce sufficient friction between the individual units to enable the composite structure to act as one rigid unit.

When construction is undertaken in this way, the various segments can be cast and cured under factory conditions, so that concrete of great strength is produced reliably, free from the worries of rain, cold, or drying winds. The segments can then be made in moulds which are machined to ensure that the mating end-faces are truly plane. The use of such high-strength factory-made concrete leads to a reduction in the concrete quantity required, which in turn achieves a saving in the dead weight of the structure, and hence leads to economy in the amount of steel. Shuttering quantities are also reduced.

A particular advantage post-tensioning has over pre-tensioning is that the cables can be formed to a curved line along the length of the member. Thus, for example, they can be low down in a beam at midspan to provide maximum reaction to the maximum imposed bending-moment, and rise nearer to the mid-depth of the section near the ends where the bending-moment effects diminish.

Post-tensioning systems

There is a variety of patent post-tensioning anchorage systems, all of which have been in successful use over many years in this country and abroad. The system illustrated in Fig. 8.4 was conceived and designed

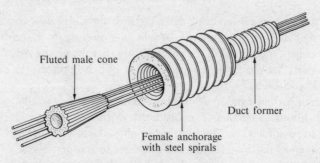

Fluted male cone

Duct former

Female anchorage
with steel spirals

FIG. 8.4. Freyssinet anchorage system.

by the French engineer Freyssinet. and was the first to be developed on any scale and is certainly the best known.

In the Freyssinet system, up to a maximum of 12 wires are grouped around a helical core and taped together to form a cable. At each end of the cable, the wires are anchored by being held between two concrete cones which fit one inside the other. The female part of the anchorage consists of a conical steel-wound lining heavily reinforced with high tensile steel spirals to resist bursting forces. The concrete male cone has a fluted surface to space evenly the requisite number of wires. The stressing of the cables is carried out by a special jack which first tensions the wires, and then, with a subsidiary ram, forces the male cone home to achieve the anchorage. Finally grout is injected to the cable within the duct through a small hole provided in the centre of the male cone.

Loss of prestress

Mention should be made of the problem of loss of prestress which arises from two main causes as follows. Firstly concrete shrinks on hardening over the first year of its life; and whilst this can be minimized by careful control of the water content, it cannot be eliminated. Secondly, when concrete is under persistent stress it suffers long-term irreversible strain known as creep, which is about three times the amount of the initial elastic strain. Accordingly over a period, part of the initial stretch of the prestressing wires is lost, and so also, therefore, is part of the initial prestress.

This is one of the reasons why in prestressed concrete work it is necessary to work to such high steel stresses and achieve such considerable stretching of the steel. Other losses arise from bond or anchorage slip; and in post-tensioned work from friction between the wires and the duct surfaces at the time of prestressing.

In pre-tensioned work the total losses may be as high as 30 per cent of the initial prestress; in post-tensioned work total losses of about 20 per cent are more normal.

Materials and workmanship

An essential aspect of prestressed concrete construction is the need for materials and standards of workmanship to be of very high quality.

Concrete for prestressed work has to be about 50 per cent stronger than that commonly used for ordinary reinforced concrete work. Typical characteristic strengths of concrete for prestressing are 40 or 50 N/mm^2 as compared with 25 or 30 N/mm^2 for ordinary reinforced concrete. This is achieved by extra careful selection and grading of the aggregates, and strict control of the water/cement ratio, which in turn makes greater demands on operatives' skills in ensuring proper compaction of the concrete.

The steel wires, which are either used singly, or wound in groups into cables, are manufactured by a process of cold-drawing; this changes the metallurgical properties of the wire close to its skin and greatly enhances its strength. As a consequence the smaller the diameter, the larger is the proportion of skin affected, and the higher the average characteristic strength of the wire. Both wire and cable are manufactured and supplied coiled in long lengths.

9

Precast concrete

Most concrete is cast *in situ* in its final position in the finished structure; however sometimes concrete units are *pre-cast* at ground or bench level, and after hardening lifted into place to form part of the structure. Precasting may be done on the same site as the finished structure, but is more usually done away at a special works. Some firms specialize entirely in the manufacture of precast units; others follow this up by doing the erection work on site as well.

The prime advantage claimed for precast concrete work is that standardized units can be mass-produced in comfort whilst the rougher tasks of foundation and other preparatory works are being carried out on site in all weathers. The precast parts are then delivered at the appropriate stage and erected quickly without the complication on site of shuttering, steel-fixing, or concrete mixing and placing. It is not essential for the units to be standardized; but the more one breaks away from simple repetition, the more one loses one of the outstanding features of the precast manufacturer's art.

Precasting works are normally located handy for the supply of good consistent concreting stone and sand; and after a period of time should be capable of producing reliably a concrete of very high standard with well understood characteristics. Furthermore precasting works are likely to attract a more permanent team of better skilled tradesmen than are found on the average outside construction site.

There are, on the other hand, a number of practical limitations on the use of precast concrete, and it is only right to balance the picture here by mentioning a few of the disadvantages and objections. One of the great merits of cast-*in-situ* concrete over its main competitor, the structural steel frame, is that it can be formed in an infinite variety of shapes and sizes without special fabrication and connection problems. And concrete, when cast *in situ*, produces *ipso facto* a monolithic rigid form with all the benefits of a continuous structure; this is a feature that can only be achieved in structural steelwork by the use of site-welded connections, which are inconvenient and expensive to produce.

In these two regards, *precast* concrete seems generally to have lost out; and whatever other points there may be in favour of precast work, these need to be weighed very carefully against the obvious drawbacks mentioned above. Nor should one forget the costs and difficulties with precast work of transport, co-ordination, storage, erection, and damage. How much damage is acceptable at the arrises of a precast member before it has to be rejected?

With precast work there are special problems too relating to tolerances affecting bearing pressures and spalling risks at all joints and connections. There is also the need to ensure that units are adequately tied together to prevent progressive collapse, particularly with multi-storey systems. These points will be clearer from the descriptions that follow of a few simple building units.

No attempt is made in this chapter to deal with portal-frames for industrial sheds, bridge-deck units, piles, or other items where precasting may have special attractions.

Floor units

The simplest precast item is the floor unit, and this is much used in all types of buildings, whether these be steel-framed, concrete-framed, or simple load-bearing brickwork. Many standard types of unit are produced, some of them pre-stressed, though not all. Fig. 9.1 illustrates two typical types; both are cast on a steel bed at floor level as a continuous length, often about 100 m long, and are subsequently cut into shorter units as required.

The narrow unit in Fig. 9.1(a) is normally cast between two fixed steel side-shutters. The central core is formed of a long 'sausage' of expanded polystyrene or similar, which serves to keep the dead weight of the unit down whilst still achieving an over-all depth for the member sufficient to avoid excessive deflection. Such a unit 400 mm wide x 150 mm deep would weigh only about 55 per cent of a solid unit, yet the strength and stiffness requirements are adequately met by the side-ribs and top-flange acting together as little L-beams.

Units of the type shown in Fig. 9.1(b) are formed by an interesting modern technique avoiding the use of all formwork other than the soffite bed. The concrete is made with relatively fine aggregate, mixed much drier than for normal work. It is spread and vibrated between top and side plates which move along as part of a special concreting machine which crawls its way on rails the length of the bed. The same machine has horizontal bullet-headed steel formers which create the

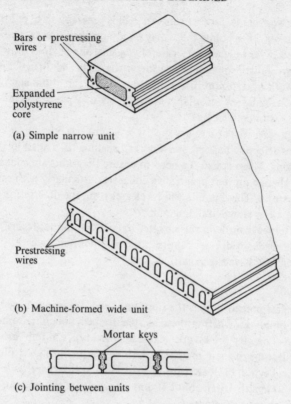

Bars or prestressing wires

Expanded polystyrene core

(a) Simple narrow unit

Prestressing wires

(b) Machine-formed wide unit

Mortar keys

(c) Jointing between units

FIG. 9.1. Precast floor-units.

continuous circular voids by reciprocating in and out of the working face of the concrete as this advances behind. In this way a single passage of the one machine leaves behind itself the concrete completely formed to the profile shown, rather like tooth-paste being released from a tube as it is drawn slowly along the top of a brush. The concreting machine is operated by one man, and travels the 100 m length of the bed in about 1½ hours. The only other operation in creating the unit is the fixing of the prestressing wires which are stretched from one end of the bed to the other before the concreting machine begins its journey.

It will be noticed that both types of unit in Fig. 9.1 have specially shaped side profiles. After the units have been laid side by side in position on the job, the space between the units is packed solid with cement/sand mortar to form a key as shown at Fig. 9.1(c). This has the effect that no unit can then deflect without bringing down with it its neighbours, so that the task of carrying a load concentrated on one unit is shared between the units on either side. Sometimes, in addition, for the same purpose, a continuous concrete topping is laid over the units, strengthened with a mesh of steel reinforcement. The proper keying together of precast floor units serves also as a prevention against floor finishes being damaged along the lines of the joints.

Structural frames

Floor units can form part of a wider range of units which together constitute a total building frame. Systems are available which include load-bearing walls, mullions, columns, beams, slabs, stair-flights, and all manner of lesser items. Generally speaking, the simpler and more repetitive the system, the greater the likelihood of success. The builders of Stonehenge had the right idea; so did the ancient Greeks. Some obvious applications today include multi-storey car-parks, commercial office blocks, and cheap medium-rise housing including flats and maisonettes. By contrast, much of the facade of the American Embassy in Grosvenor Square was obviously precast.

Any precast structure has to be designed taking into account the fact that the joints will be less rigid than we get with *in-situ* concrete construction. Individual units are usually assumed to be non-continuous and to have pinned ends. Over-all stability has therefore to be achieved by the use of walls, whether precast or otherwise. In tall buildings the stiffness is normall provided by the *in-situ* central core which contains the lifts, escape stairs, lavatories, and services.

Fig. 9.2 is an 'exploded' illustration of some typical connections between precast structural units. Fig. 9.3 enhances the same information, though in practice exact details can vary considerably. In the method shown, the supporting column has a large diameter steel bar projecting up above its top, and this goes through and beyond a vertical hole left in the beam. The beam is packed off the column at its correct level with steel packs, leaving a clear gap of about 25 mm; polystyrene gaskets are then fitted between the packs, and a cement/sand grout of porridge consistency is poured down the hole to fill the gap and also the hole. After the grout has set, the packs and gaskets are removed, and the joint is completed by ramming in a dry-pack mortar having the least amount of water that will just cause the sand particles to bind together.

Later, when it comes to connecting the upper column, the detail is

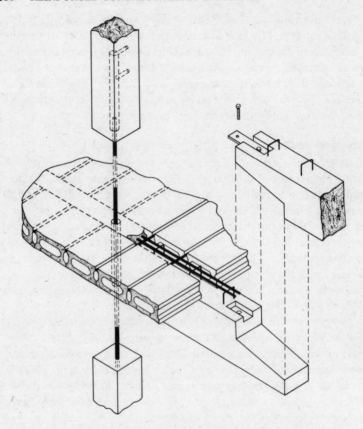

FIG. 9.2. Structural connections: (exploded illustration).

very similar, only this time the same bar is received into a hole left in
the bottom of the column. Two grout holes are provided in one of the
sides of the upper column, one for the grout to be poured in from a
little feeder tank held 2 m above, and the other to enable air to escape
freely until filling of the joint has been completed.

The beam-to-beam connection is arranged at the position where the
point of contraflexure (zero bending moment) would have been, had
the structure been continuous. Thus some moment effect is achieved at
the column by cantilever action; and the midspan part of the beam is
shorter and therefore subjected to less bending. With this arrangement
the joint is required to transfer no moment (it couldn't anyhow); and
the shear to be transferred is considerably less than the maximum
which occurs where the beam reaches the column. The bearing surfaces

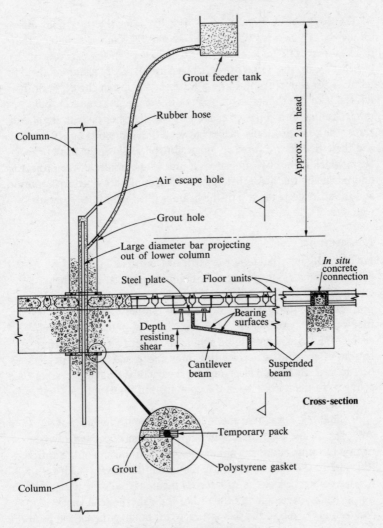

FIG. 9.3. Structural details.

at the joint are inclined so as to enhance the depth of beam available on both sides for taking the shear. The sequence of operations in forming the joint is as follows. The midspan part of the beam is supported on props; the steel connecting plate holds the beam steady through the temporary stages; dry-pack mortar is rammed in to fill the joint; and finally the props are removed.

The slab-to-beam connection is made by laying the floor units direct on the beam with a seating of about 75 mm. This shows strikingly the need for manufacture and erection to be to very close tolerances; if the bearing, by accumulation of errors, became markedly less there would be cause for concern; and if the surfaces of the top of the beam or the underside of floor units were rough or uneven, point contact would result, leading to real risk of spalling. Small diameter tie-bars are placed across the beam from the spaces between the floor units on either side; and these tie-bars are wired to longitudinal bars along the top of the beam which themselves are wired in turn to the insides of the projecting hoops. Finally the spaces between the floor units are filled with mortar, and the strip along the top of the beam is filled with fine-aggregate concrete.

Non-structural units

Precast units are often used for partitions and cladding, but for lightness's sake are manufactured to dimensions that make them unsuitable for any load-bearing duties. The variety of shapes and profiles for cladding panels seems almost limitless; however one feature they generally have in common is their floor-to-floor height which enables the fixing details to be simple and repetitive. For practical reasons of erection, waterproofing, movement, and appearance, it is best to keep cladding panels out at the edge of the building, rather than trying to fit them in between the columns and floors. Fig. 9.4 indicates a typical unit.

It should be noted that the unit has its weight supported at one level only, in this case the top; this allows for movements of the unit (due mainly to temperature and shrinkage), and movements of the structure (due also to floor loading and foundation settlements). The load-bearing connection to the structure is the mortar joint between the nib on the back of the unit and the top of the slab; the steel angle connecting the nib to the slab is merely for holding the unit in position. At the foot of the unit, a clear gap is left above the unit underneath; and the connecting angle here is slotted so relative vertical movement can take place freely. On important buildings metal fixings of this type are either stainless steel or phosphor-bronze to avoid corrosion from condensation causing damage to the fixings and thus to the units.

Fig. 9.5 shows two widely different techniques for achieving watertightness of the joints at the edges of the cladding panels. The 'sealed' joints at (a) make a frontal resistance to the entry of water at

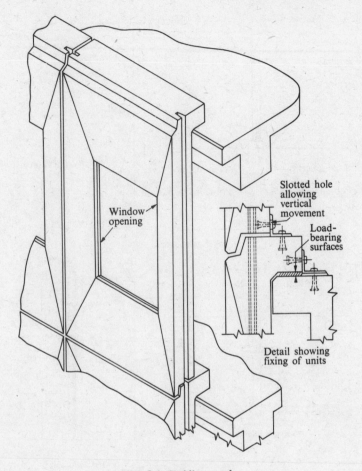

FIG. 9.4. Cladding-panel.

the weather face, where the task is probably most difficult; the 'open-drained' joints allow the water into the joint and there effect the necessary arrest under more favourable conditions, allowing the water subsequently to drain away.

The 'sealed' joint is commonly formed with a polysulphide sealer applied by gun against a foam backing-strip previously squeezed into the gap. Polysulphide has the virtues of adhering well to the surface of the concrete and retaining its flexible nature over a long life. In addition an adhesive plastic strip is normally squeezed into the gap at

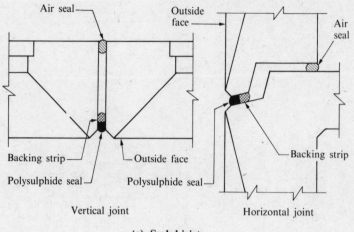

(a) Sealed joints

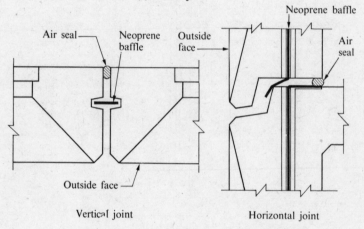

(b) Open drained joints

FIG. 9.5. Joints between cladding-panels.

the inner face, partly as an air-seal but also serving as a second line of defence.

In the 'open-drained' type of joint, the vital defence is the plastic air-seal strip at the inner face; but loose neoprene baffles in the vertical joints effectively prevent the water getting to the seal, and the sloping profile of the horizontal joints serves the same purpose. Generally the open-type joints are the least offensive aesthetically.

Reference has been made earlier to the need in precast work to adhere strictly to close tolerances. This applies very much with cladding panels, where variations in joint widths can be recognised easily by even the most untrained eye. It is very difficult in manufacture to get the panel widths consistent within 5 mm of one another, yet 5 mm is proportionally a large variation in the width of joints intended to be say only 20 mm wide. The 'sealed' joints seem rather to accentuate the variations, whereas the 'open-drained' joints help to lose them.

Special finishes

Precast manufacturers have shown much ingenuity in developing a variety of techniques for producing surface effects of different colours and textures. A few of these are as follows.

One is to grind the surface of the finished concrete so as to remove the rough outer cementy layer to a depth of about 2 mm and leave a smooth polished face. The thickness that is ground away will include part of some of the large pieces of aggregate and these pieces appear in cross section in the polished surface. Care in selecting the aggregate and close attention in pouring the concrete into the moulds are of course essential if a pleasing and uniform appearance is to be obtained. The use together of light coloured limestone aggregate and white cement, available as an alternative to the normal grey cement, though at some additional cost, is a combination much in vogue.

Another method giving an exposed aggregate finish is to paint the relevant parts of the formwork with a retarding agent just before pouring the concrete. This has the effect of slowing the rate of hardening of the cement at the surface in contact with the formwork, so that when the unit is removed from its mould, about a day after casting, the cement matrix in the surface zone can be removed by gentle brushing under a stream of water. With careful control of the amount of retarder used, a rough textured surface can be obtained in which the large aggregate particles stand proud of the remaining matrix by 5 mm or more.

A variation on this for exposing larger aggregates, say 30 mm or over, is the sand-bed technique, where the selected face-aggregate is placed firmly by hand in a layer of dry sand previously spread in the bottom of the mould. The aggregate is then covered with a workable mortar, and finally the mould is filled with concrete using hand tools taking great care to avoid causing disturbance. When the unit has hardened and been removed from its mould, the sand is simply brushed off the face to expose the aggregate.

Another tecnnique, known as the aggregate-transfer method, entails sticking the chosen aggregate onto plywood or hardboard with adhesive, and then fitting the plywood as a lining inside the mould. The unit is then cast in the normal way; and after the mould has been removed the plywood is stripped off, leaving the aggregate exposed. Any adhesive remaining is cleaned off subsequently with a wire brush.

There are many other ideas being used and developed; and more will come. Perhaps next time the reader is waiting in a bus queue or a traffic jam he will be tempted to take a critical look at one or two concrete panels in the vicinity and see if he can work out how the finish of these was achieved.

10

Water-retaining structures

Introduction

Water-retaining tanks, reservoirs, and water-towers have a special requirement beyond those of other concrete structures, namely that they must not leak. To this end additional precautions have to be taken against cracking and percolation.

In a normal reinforced-concrete structure, the very fine cracks which occur in the concrete, whether as a result of stressing under normal design load, or whether due to the natural shrinkage of the concrete on setting and drying out, or whether the result of tensions set up as a result of cooling of the structure after a warmer period, are usually of no significance. In the case of water-retaining structures, however, these cracks may lead to leakage which is unacceptable, and which in any case would be disfiguring and likely to be the cause of corrosion.

In the chapter on prestressed concrete we saw how it was possible by prestress to ensure that the whole concrete section remains in compression even though the member be subjected to bending or even direct tension. Thus we have in prestressed concrete one satisfactory method of avoiding tensile stresses and cracking of the concrete, and hence one solution as a basis for the design of liquid-retaining structures, provided of course the structures are themselves supported in a manner sufficiently free for the prestress to be able to move the concrete together so it is in fact everywhere in compression. This would not be the case where a structure is built in considerable lengths along the ground, and the friction forces between the ground and the structure are greater than the prestress force can overcome and exceed.

Other techniques are available for overcoming the difficulties of concrete cracking which do not suffer the limitations described above for prestressed concrete, and in this chapter we shall be devoting our attention to one of these methods.

The subject is covered in detail in the British Standard B.S.5337 (1976) *Code of Practice for the Structural Use of Concrete for Retaining Aqueous Liquids.* This Code first deals with appropriate design

117

precautions in accordance with the limit-state philosophy, the basis of which has already been described in the preceding chapters of the present book. It then goes on to give alternative recommendations related to what is known as the *elastic theory*; and as, at the time of writing, there is no background experience to support the limit-state method as applying to water-retaining structures, whereas we know from several decades' successful results that the elastic theory is certainly reliable, we shall indicate here the basis of the well-known

Elastic theory

In the elastic method of design, instead of considering *ultimate* conditions as we did with the limit-state method, we focus our attention on the *working* conditions of the structure at the stress conditions that actually occur in normal performance. Thus we do not apply a partial safety factor to our loads (which certainly seems eminently appropriate when dealing with water, the unit density of which we know precisely); but instead we apply a larger factor of safety to the material strengths, and thus arrive at what we call *permissible stresses*.

Strength calculation

For security reasons we first calculate that the structure is strong enough. Nevertheless we work to modest stresses in the reinforcing steel, so that even if a few cracks do occur (despite our attempts described later to avoid them) they are likely to be of limited width. For parts of the structure exposed to a moist atmosphere or subject to alternative wetting or drying, the permissible stress in the steel is 85 N/mm^2. Where continuous contact with the water can be assured, a stress of 115 N/mm^2 is permitted; and further increases of about 15 per cent are permitted where deformed bars are used. Nevertheless all the values come well within the competence of ordinary mild steel; and clearly in these situations where our main concern is to limit the *stretch* of the steel, there is no purpose in using anything stronger, since, as we saw in Chapter 3, the elastic modulus (E_s) for steel is 200 kN/mm^2 whether it be mild or high-yield.

We shall now follow the arguments of the elastic theory as applied to a reinforced concrete member subject to bending. The underlying assumptions are:

(a)   that the steel is an elastic material with E_s equal to 200 kN/mm^2, and that the concrete is an elastic material having an average modulus (E_c) equal to 13 kN/mm^2, so that the *modular ratio* is given by

$$m = \frac{E_s}{E_c} = \frac{200}{13} = 15 \text{ (see equation (3.3))};$$

(b)   that in members bending within the elastic range, the compressive and tensile stresses vary directly as the distance from the neutral axis (This follows from Hooke's Law, and by remembering from our beam in Fig. 3.1 that the total amounts of shortening and lengthening of the ends of the beam vary directly as the distance from the neutral axis);

(c)   that the concrete has cracked in tension up to the position of the neutral axis, so that the whole tension is carried by the steel alone.

Let us consider a beam or slab in bending. In this case it will be simplest to avoid algebraic symbols as much as possible, and adopt immediately the permissible stresses we propose to work to, namely

for Grade 25 concrete in bending = 9·15 N/mm^2
for plain steel bars in tension = 85 N/mm^2.

First we must find the position of the neutral axis, which we do as follows. The shortening of the concrete is proportional to 9·15 to some scale; and since the steel stretches only one-fifteenth as much as the concrete for the same stress $(m = 15)$, the extension of the steel will be proportional to 85/15 = 5·65 on the same scale as that which gave 9·15 as the shortening of the concrete. Hence, if to any scale (see Fig. 10.1) we set along 9·15 units to one side of a vertical line at the top to represent the shortening of the concrete, and set along 5·65 units to the other side at the steel line to represent the extension of the steel, the oblique line joining these two points represents the strain at any distances down in the beam or slab, and where it cuts the vertical basis line gives the position of the neutral axis. If this be done graphically, it will be found that the depth of the neutral axis is

$$d_c = 0·62\, d,$$

where d is the effective depth to the tension steel. This is always true with the permissible stresses we have adopted.

Now if the breadth of the beam or slab is b, the area of the compression zone above the neutral axis is 0·62 bd; and with our

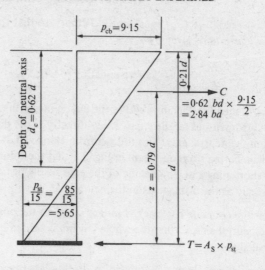

FIG. 10.1. Bending in slab or beam. (Elastic theory for strength calculation).

elastic triangular stress distribution the average compressive stress will be half 9·15, so we have the total compressive force

$$C = \text{area} \times \text{stress} = 0·62 \ bd \times \frac{9·15}{2} = 2·84 \ bd \ \text{(N)}.$$

And clearly, from Fig. 10.1, the centre of this compressive force acts at the centre of gravity of the top triangle representing the stress in the concrete, i.e. at one-third of d_c from the top of the section. This is $d_c/3 = 0·62 \ d/3 = 0·21 \ d$. The resistance arm of the section, the distance separating the compressive force C and the tensile force T, is therefore

$$z = d - 0·21 \ d = 0·79 \ d.$$

The moment of resistance of the section, as limited by the compressive stress in the concrete, is then

$$M = C \times z$$
$$= 2·84 \ bd \times 0·79 \ d$$
$$= 2·24 \ bd^2 \ \text{(N mm)}. \tag{10.1}$$

The reader will recognise a similarity between this and equations (4.6) and (4.7) which, for the same Grade 25 concrete, gave

$$M_u = 3 \cdot 75 \, bd^2 \text{ (N mm)}.$$

The greater part of the difference is due to equation (4.6) being based on ultimate conditions, relying of course on a partial factor of safety of about $1 \cdot 5$ being applied to the characteristic loads; the remainder of the difference arises from the different stress distributions assumed as between Fig. 10.1 and Fig. 4.4.

So far as the tension stress in the steel is concerned, following the elastic theory (as shown in Fig. 10.1) we have

$$M = T \times z$$
$$= (85 \times A_s) \times (0 \cdot 79 \, d)$$
$$= 67 \, A_s d \text{ (N mm)}. \qquad (10.2)$$

This compares with equation (4.5), which, for mild steel being stressed normally to 250 N/mm^2, when d_c is half d, gives

$$M_u = 163 \, A_s d.$$

Here (as with the concrete limitation) part of the difference arises from the margin provided for the application of the partial factor of safety to the loads; but this time nearly half the difference comes from the use of the deliberately low permissible steel stress we have adopted of 85 N/mm^2 so as to reduce the width of any cracks that may occur.

Calculation for resistance to cracking

In addition to checking that the structure is strong enough (albeit using modest stresses and hence low stretching of the steel), we also do a calculation to satisfy ourselves that the dimensions of the section are such that the concrete is unlikely to crack at all. With concrete being brittle, and to that extent unreliable in taking tensile stresses, the results of such calculations are obviously associated with a fair amount of chance; but we have already just shown that in terms of structural safety we have very adequate margins.

For members in direct tension we limit the concrete stress to 1·31 N/mm^2, and for members in bending we limit the tensile stress to 1·84 N/mm^2. The difference between the two is because in bending we assume (under the elastic theory) a triangular distribution of stress which in difficulty could relieve itself somewhat towards the profile shown in Fig. 4.3(a). The significance of tensile stresses at or below 2 N/mm² will be clear to the reader from what was said in Chapter 3; our margins against cracking are clearly pretty small.

Fig. 10.2 shows the case of a plain concrete member (unreinforced) subject to bending. The neutral axis is exactly at the mid-point of the section, so that the maximum stresses due to compression and tension are equal, in our case 1·84 N/mm².

Therefore $C = T$ = area × stress

$$= \frac{bh}{2} \times \frac{1·84}{2} = \frac{1·84}{4} bh \text{ (N)}.$$

The distance between C and T is $2/3\ h$.

Therefore the moment of resistance of the section is

$$M = \left(\frac{1·84}{4} bh\right) \times \frac{2}{3} h$$

$$= 1·84 \frac{bh^2}{6}$$

$$= 0·31\ bh^2 \text{ (N mm)}. \tag{10.3}$$

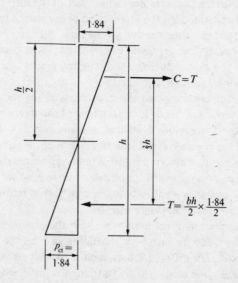

FIG. 10.2. Bending in slab or beam. (Elastic theory for resistance to cracking).

This is obviously a much lower result than we had at equation (10.1), and shows the great significance of the calculation for resistance to cracking, which in most situations leads to considerably thicker sections than would be necessary for normal structures not required to be water-retaining.

It is true that equation (10.3) ignores the effect of any reinforcement, but for the uncracked section, with $m = 15$, the tensile stress is limited in the case of direct tension to 19.6 N/mm^2, and for bending to about 27.6 N/mm^2, so the error of ignoring the steel is only of the order of 20 per cent or so.

Shear and bond stresses

The foregoing explanation of the elastic theory, particularly as applied to water-retaining structures, is accurate as far as it goes but nevertheless not complete. In particular any reader wishing to study the matter further should be on the look-out for differences over the matters of shear and bond.

The permissible shear stress calculated on the basis of V/bz is limited to 0.77 N/mm^2 for Grade 25 concrete, and once that figure is exceeded it has to be assumed the concrete has cracked and can make no contribution whatever to carrying inclined tensile stress. Accordingly provision has to be made thereafter for the whole of the inclined tensile force to be carried by the steel reinforcements alone. A further difference arises as to the maximum compressive shear stress permitted, which is limited to 1.94 N/mm^2 on the reduced area.

It is also wise in water-retaining structures to minimize bond stresses. This is because of the reduced bond which develops where reinforcements are in saturated concrete, and results from the concrete shrinking less and consequently gripping the bars less tightly. Bond stresses are referred to further in Chapter 11.

Shrinkage and temperature conditions

Beyond the calculations described above in accordance with the elastic theory, further preventive measures have to be taken against cracks forming due to the *shrinkage* of the concrete and the *cooling* of the structure. The first of these is the provision of a *minimum percentage of reinforcement* of 0.5 per cent in both directions at right angles where plain bars are used, and 0.3 per cent when deformed bars are used. Where slabs are 200 mm or over, these amounts of reinforcement have to be shared equally by placing half the quantity in each face.

A further precaution against the formation of shrinkage and cooling cracks is made by limiting the length of work concreted in any one pour to 7·5 m. Special precautions are then taken to seal the cracks which will form at the ends of these 7·5 m operations. Indeed the matter of forming and sealing joints in water-retaining structures is so important that it is worthwhile now giving special attention to the point.

Joints in construction

Construction-joints are left at right angles to the surface of the slab with a carefully formed rebate (tongue and groove fashion), the width of rebate occupying approximately one-third the thickness of the slab (see Fig. 10.3(a)). The surface of the old concrete is carefully cleaned by hacking and watering, and a layer of cement-sand mortar placed on the hacked surface immediately before further concreting. With construction-joints, the reinforcements are continued through the joint. A seal is normally provided on the water face.

Partial contraction-joints are spaced at intervals of about 7·5 m in order to allow the concrete to shrink freely on setting and hardening. In this way cracks are prevented from forming intermediately between the joints. Partial contraction-joints are normally formed without any rebate (see Fig. 10.3(b)), and the surface of the old work is not hacked, as the desire is to avoid adhesion between the old and new concrete; indeed this is the point where movement or separation is deliberately encouraged. At partial contraction-joints, the reinforcements are left running through the joint. A waterstop is normally provided, preferably centrally in a wall, or on the undersurface of a floor as illustrated. A seal is essential on the water face.

Sometimes *complete contraction-joints* are required, in which case the reinforcements are stopped at either side of the joint in order that there may be complete freedom for the joint (concrete and steel) to open up. The general form of such joints is indicated in Fig. 10.3(c). There is a point of view that complete contraction joints are as much trouble to form as expansion joints, and since the latter provide a double function, some engineers have given up the practice of providing complete contraction-joints, and use expansion joints instead.

Expansion-joints are required at intervals of about 30 m. In order to allow the joint to close partly when the concrete on either side expands due to temperature or other causes, a flexible filler is built into the joint as shown at Fig. 10.3(d). The filler has to be readily compressible, and at the same time capable of recovery when the joint ceases to be

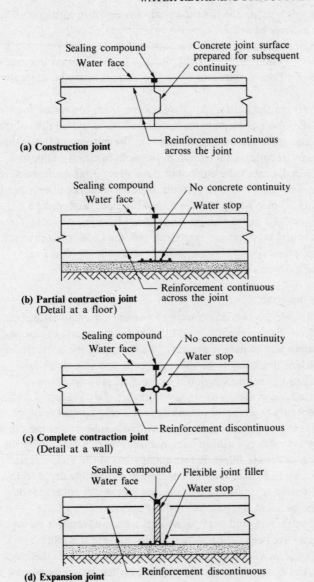

(a) Construction joint

(b) Partial contraction joint
(Detail at a floor)

(c) Complete contraction joint
(Detail at a wall)

(d) Expansion joint
(Detail at a floor)

FIG. 10.3. Joints in construction.

under compression. The filler material can be of non-rotting cork base, or alternatively of cellular cane-fibres impregnated with a bitumen composition to give protection against rot. With expansion joints a waterstop is essential, preferably centrally in a wall, or on the under-surface of a floor as illustrated. A seal is also necessary on the water face.

Sometimes the walls of circular tanks or water towers are founded on a *sliding-joint* of bituminous material which is laid on the concrete floor slab before the walls are concreted. The purpose of this is that when the circumferential tension in the walls causes the tank to increase in diameter, the bottom of the wall is free from restraint and can move outwards. If such restraint is not avoided, cantilever bending moments develop between the wall and the floor; and whilst a satisfactory design can then be made here with adequate concrete thickness and reinforcements to cope with the bending moments, it is often satisfactory and cheaper to provide the sliding joint so that bending moments do not occur in this position.

Sealing materials

At all flexible joints it is necessary to provide a mechanical *seal* on the water face of the concrete. In addition to being watertight, the sealing compounds also require the quality of adhering to the concrete, as well as sufficient flexibility to be able to stretch or compress without cracking or excessive extrusion. For practical purposes these compounds need also to be readily workable. Two groups of materials have generally been found suitable; one is a cold-applied two-component compound based on polysulphide rubber, and the other is a hot-applied rubber—bitumen compound sometimes built up by the inclusion of certain fillers. Before application of the sealing materials, it is necessary to be sure that the concrete is completely dry, and then a special primer is applied to the concrete face which enhances the adhesion between the concrete and the sealer.

At contraction- and expansion-joints where movements of the structure have been deliberately collected and are therefore accentuated, *continuous flexible waterstops* are cast into the concrete and extend across the joints as described earlier. Where movement is not likely to be great, for example at partial contraction-joints, these waterstops can be of polyvinyl chloride (pvc). Where greater movements are expected, particularly at expansion-joints, the waterstops are normally of synthetic rubber. Where the waterstops come within the centre of the concrete member, they are made of dumb-bell cross-section, so that the more they are pulled, the tighter the end bulbs bear against the concrete; and where the waterstops come on the face of the concrete (notably where the concrete is poured against the ground), the special former as illustrated is used with upstanding ribs of which there should never be less than two to each side of the joint.

Precautions against percolation

To guard against percolation, it is important that close attention be given to the design of the concrete mix. With this proviso, our Grade 25 concrete is adequate for the purpose. A minimum cement content of 360 kg/m^3 is desirable to ensure good density and workability; on the other hand the maximum cement content should never exceed 400 kg/m^3 so as to avoid risk of excessive shrinking and formation of cracks on setting and drying.

A useful rule-of-thumb precaution against percolation is to provide a minimum thickness of reinforced concrete slabs of 25 mm + 1/40th the depth below water level; and a minimum value of 100 mm. Thus the thickness of a floor slab for a reservoir 6 m deep below water level would need to be

$$25 + \frac{6 \times 10^3}{40} = 25 + 150 = 175 \text{ mm.}$$

Similarly the thickness required 3 m below water level would be

$$25 + \frac{3 \times 10^3}{40} = 25 + 75 = 100 \text{ mm;}$$

and at any lesser depth below water level the thickness should not be reduced below the 100 mm value.

This requirement of minimum thickness is intended to reduce the hydraulic gradient through the slab thickness so that with good materials and workmanship the concrete will be such as to give practical watertightness.

Examples of structures

For concrete tanks of anything beyond the most modest proportions sitting directly on the ground, the circular plan indicated in Fig. 10.4(a) is much the cheapest. The water presses radially against the walls with

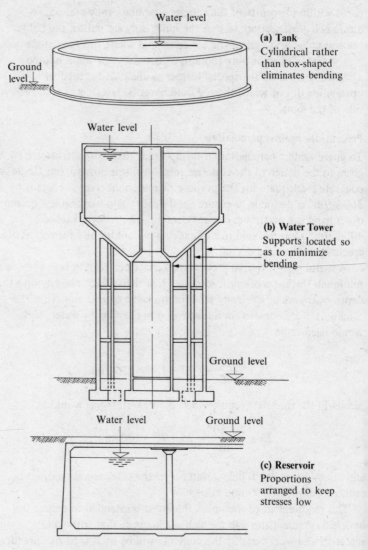

Water level

(a) Tank
Cylindrical rather
than box-shaped
eliminates bending

Ground
level

Water level

(b) Water Tower
Supports located so
as to minimize
bending

Ground level

Water level   Ground level

(c) Reservoir
Proportions
arranged to keep
stresses low

FIG. 10.4. Examples of water-retaining structures.

equal intensity at any depth, so that the walls are subjected only to
pure hoop tension, with no problems of bending. It is true that the
circular formwork is somewhat more expensive than for a straight-sided
tank, but the virtues of hoop tension are so considerable that for all but

the very smallest tanks this outweighs the additional cost of the shuttering. As an example we shall consider a tank 25 m diameter at a depth of 5 m. The hydrostatic pressure is therefore $5 \times 10 = 50$ kN/m^2, and the circumferential or hoop tension acting across any diameter trying to burst the tank in half is

$$T = 50 \text{ kN/m}^2 \times \frac{25 \text{ m}}{2} = 625 \text{ kN per metre height of wall.}$$

Considering first the matter of strength, we calculate the area of hoop steel required to hold the tank together, working to the permissible stress of 85 N/mm^2. This is

$$A_s = \frac{625 \times 10^3 \text{ N}}{85 \text{ N/mm}^2} = 7350 \text{ mm}^2 \text{ per metre,}$$

and provided by 25 mm bars at 125 mm centres in both faces

(area of $2 \times 3930 = 7860$ mm^2 per metre).

Next we choose a suitable thickness for the wall, and check that the tensile stress in the concrete is kept down below the permissible 1·31 N/mm^2. We will try a wall 375 mm thick, for which the equivalent concrete area per metre is

$$A_E = (375 \times 1000) + (14 \times 7860),$$

where 14 is the modular ratio (m) less unity for the area in the wall actually occupied by the steel. Thus

$$A_E = 485\,000 \text{ mm}^2 \text{ per metre,}$$

and the concrete tensile stress is

$$f_{ct} = \frac{625 \times 10^3 \text{ N}}{485 \times 10^3 \text{ mm}^2} = 1 \cdot 29 \text{ N/mm}^2,$$

which is satisfactory.

A check against percolation shows we need a minimum thickness of

$$25 \text{ mm} + \frac{5000}{40} \text{ mm} = 150 \text{ mm,}$$

which is well taken care of by the thickness of wall already chosen.

Lastly, for vertical reinforcements, we shall require at least the minimum of 0·25 per cent in both faces (if we use plain bars), giving us

$$\frac{0 \cdot 25}{100} \times 375 \times 1000 = 940 \text{ mm}^2,$$

which is provided by 16 mm bars at 200 crs, giving an area of 1010 mm^2 per face per metre.

A sliding-joint can be provided at the bottom of the wall; alternatively there can be rigid continuity between the wall and the floor, in which case suitable cantilever reinforcing steel will have to be provided.

The water tower in Fig. 10.4(b) similarly relies on the principle of hoop tension, both in the external walls and the outer part of the sloping floor. The access shaft up through the middle is likewise free from bending effects because of the equal radial pressures imposed upon it inwards from the water, causing a state of ring compression. The supports to the underside of the water-containing compartment are carefully disposed so that the bending effects on the floor are generally in balance, and therefore minimized. There are many spectacular variations to solutions of this type which engineers and architects have jointly been at pains to exploit.

Fig. 10.4(c) shows part of a service reservoir. Here the overall size and configuration requirements are such that a circular solution would not be appropriate. In such circumstances the spacing of columns and the proportions of the members have to be arrived at with great care to achieve an optimum solution giving the highest possible storage capacity for the least amount of bending, and hence the minimum unit costs of materials and temporary works.

11

The anatomy of concrete

So that we could make a start on discussing the general theory of reinforced concrete design in very simple terms, we described in Chapter 1 the basic physical properties of steel and concrete: and without going into any great detail we settled for a Concrete Grade 25 and said a suitable *prescribed mix* that would give the 25 N/mm² strength would be:

cement	50 kg,
sand	85 kg,
stone	160 kg,
water	28 kg.

What is the explanation of these proportions? How much do they matter?

The first requirement of good concrete is that it should be solid right through, without porosity, voids, or 'honeycombing'. It will then be as dense and strong and permanent as the individual ingredients make possible.

The stone, which is known also as the coarse aggregate, is often a natural gravel or shingle, or can be a crushed rock such as limestone or granite. These all have somewhat different properties, as we shall see, but when standing as a heap all comprise solid stone particles with spaces or voids in between. For the concrete used in most normal reinforced-concrete work, the particle sizes of the stone vary from 20 mm to 5 mm; and depending on how the particles are graded between these limits the void content of the stone heap will be of the order of 40 per cent. The sand, on the other hand, which is often known as fine aggregate, generally consists of particles about 5 mm and less, and when standing as a heap has a void content of about 30 per cent, although, because the sand particles are much smaller than the stone, the voids in the sand are not as obvious.

In a good concrete, the sand will fill the voids in the stone, and the cement will fill the voids in the sand. Thus if the stones contain 40 per cent voids, then if the sand could be inserted between the stone particles without separating them, obviously the sand volume should be 40 per cent of the stone volume (both volumes being measured to include solid and void in natural proportion). In practice, it is impossible to get the sand to fill the voids in the stone without some sand particles getting caught between the stones where they would otherwise touch, and generally it is necessary as a result of this to add about 10 per cent (depending chiefly on the size of the particles) more than the initial void content. Thus a stone with 40 per cent void needs in practice 40 per cent + 10 per cent = 50 per cent sand.

The same applies to the sand voids. If the sand contains 30 per cent voids then if the cement could be made to fill the voids without separating the sand particles, 30 per cent of the sand volume should be the cement content. In practice, a greater volume is required, because the cement, besides filling the voids, also forms an adhesive film round the surface of sand and stone particles, and in practice about 20 per cent more cement is required, so that roughly 30 per cent + 20 per cent = 50 per cent of the sand is the cement volume required.

It will be seen therefore that if the stone needs 50 per cent sand, and the sand needs 50 per cent cement to fill all the voids, then a concrete of 4 parts stone, 2 parts sand, 1 part cement meets the requirements of a dense solid concrete. However, if the stone contains more voids, more sand and cement are needed; and if the sand contains more voids, more cement is needed. In our prescribed mix, it will be seen the ratio of stone to sand is about 1·9, and the ratio of sand to cement is about 1·7. The reason the stone to sand ratio is a little less than 2·0 (actually 1·9) is to allow a slight margin to cater for stones which have a greater voids ratio arising from the grading or shape of the particles. And a sand to cement ratio of 2·0 would have been perfectly satisfactory to achieve a dense concrete, but the Grade would then have been only 20 N/mm². The 1·7 sand to cement ratio (more cement) is what gives our Grade 25 concrete its extra strength.

The reader may then well ask why, if a sand to cement ratio of 2·0 will adequately fill the voids, a lower ratio such as 1·7 (more cement) should be desirable and advantageous to achieve a concrete grade of 25 N/mm². The explanation is that when a concrete member actually fails in compression, the material finally comes asunder due to the individual particles spalling apart in tension. This can be seen in the crushing of a 150 mm cube, which takes place by a wedge-like action of the stones splitting the cube laterally, dividing it up by vertical lines into several prisms liable to buckle. This action is resisted only by the *tensile* strength of the concrete. Now the tensile strength preventing fracture consists partly of the contact adhesion between stone and cement, and partly of the tensile strength direct between the much smaller particles of cement to each other in the interstices between the stones. The greater the cement content, the greater the proportion of area at any incipient splitting plane is cement and not stone, and therefore the greater the tensile strength.

The crushing strength of concrete is thus really limited by the tensile strength, and this is why there is an almost constant ratio between the compressive and tensile strengths, which in the case of concrete is about ten to one. By increasing the cement content beyond the minimum necessary to just fill the voids within the stone plus sand, very much greater concrete strengths can be achieved, as for example up to about 60 N/mm^2 for special precast and prestressed concrete work, and sometimes even beyond this, provided of course the strength of the aggregates is not exceeded. But this then brings us far into the range of *designed mixes* as was referred to in Chapter 1.

So far we have not mentioned the quantity of water. Sufficient water has to be added (a) to set off the chemical reaction which causes the cement later to set and harden, and (b) to lubricate the aggregate particles so they can be properly compacted to ensure a dense mass completely free of voids. However a surplus of water makes for a very weak concrete as will be explained later: indeed the strength of concrete depends more than anything else on the ratio between the weights of water and cement used in the mix, the higher the proportion of water the weaker the mix, and the higher the proportion of cement the stronger the mix. As the cement and the water both assist in lubricating the aggregate particles so the concrete can be properly compacted, it is a matter of some judgement to know how much lubrication should be provided by the use of cement, and how much by the use of water. Clearly cement is expensive whilst water is cheap: yet cement is useless if the chemical processes of setting and hardening of every particle is not initiated and seen through to completion by the presence of a sufficiency of water. The 28 kg of water described for our prescribed 25 N/mm^2 mix is approximately right for hydrating the cement and giving reasonable assistance for compaction of the particles in most ordinary situations, whilst yet not leading to excess as would make for a sloppy weak mix. It represents a water/cement ratio of about 0·55. However the exact amount of water required is much dependent on the moisture content of the sand.

If a box containing dry sand has water poured on it, the sand will settle so that it no longer fills the box. If the wetted sand is then tipped out and put back again, it will be found to be more than is required to fill the box. In the first case, the effect of water was to reduce the friction between the particles and cause them to settle down closer together, whilst in the second case, where the water is not actually poured on, but only the water adhering to the surface is left, the surface tension prevents movements of the particles, just as moist sand can be moulded while dry sand refuses to stand. For the purpose of assessing the voids, it is the densest condition we are concerned with, because in ordinary concrete there is always enough water to allow of free movement of the particles. The sand from the bottom of a heap is generally wetter than the sand from the top, and certainly more consistent.

A useful practical test for controlling on site the amount of water required to the mix is made as follows. An open-ended metal former, being a frustrum of a cone 300 mm high, 200 mm bottom diameter, and 100 mm top diameter, is stood with its base on a steel plate and filled with the concrete to be tested, the concrete being compacted by tamping in layers with a steel rod. (See Fig. 11.1). The cone is then lifted carefully upwards, so that the concrete slumps under its own weight. The amount the concrete reduces in height from the original

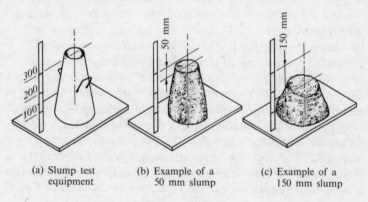

(a) Slump test   (b) Example of a   (c) Example of a
    equipment   50 mm slump   150 mm slump

FIG. 11.1. Slump-test.

300 mm is known as the *slump* of the concrete. Any slump more than 150 mm is definitely too wet, and usually a 50 mm slump is suitable for reinforced concrete under normal conditions where mechanical vibrators are being used. In cases of large members which contain only little reinforcement, a slump approaching nil may be satisfactory.

The solid material of most common stones weighs approximately 2600 kg/m^3 (though limestone may be less, and granite may be more). A good and ready test of the success attending the proportioning of concrete is clearly the weight of a 150 mm cube, since the difference between its density and the value of 2600 indicates with most materials the amount of voids in the mass left unfilled. Experiments have clearly demonstrated that the heavier cube is also the stronger. When concrete is made with a low water content and mechanical vibration a density of 2500 kg/m^3 can often be achieved, showing a voids ratio in the finished concrete of less than 5 per cent.

It is of course impossible in practice to make a concrete entirely without voids, because at the stage when the cement, aggregate, and water are being mixed together, the water represents a very real volume: and although a part of this water enters into the chemical combination with the cement, a part does not, but subsequently evaporates and is replaced by air. It becomes clear from this that an *excess of water* in concrete reduces the strength because the concrete will be less dense. But whilst an excess of water weakens the concrete by leaving water-voids which evaporate and produce air-voids, the loss of strength from excess of wetness is nothing like the loss of strength which results from *insufficiency of water.*

When there is insufficient water, a lack of density also results: the sand and cement particles, as mentioned previously, are inadequately lubricated with the effect that they cannot ease themselves into the voids between the stone particles: thus the mass does not achieve the optimum intimate contact. A second and more serious defect arising from insufficient water is that there may be not enough water to complete the chemical transformations of the cement, with the result that the cement remains partly unset, and the concrete stays relatively powdery and weak.

However much water is first mixed in with the other ingredients, the cement particles are never initially wetted right through, owing to the large surface area they present in relation to their minute weight. Over a period of weeks and months however if there is the appropriate amount of free water in the mix, this by degrees will work its way through into the core of each individual cement particle, in this way completing the hardening process and so increasing the strength of the concrete. But if at any stage in the hardening process the concrete is once allowed to dry out, a hard crust forms at the surfaces of the individual cement particles for as far as they have already been wetted, and this crust prevents further chemical action taking place. For these reasons it is of great importance to prevent concrete drying out in the early stages of its life; and generally for the first four days or so the concrete should be kept covered with an absorbent material well soaked, or sprayed with a temporary impermeable membrane which will retain the concrete in a wet condition. This is particularly so in hot weather or drying winds. After the four days the free water still remaining in the voids within the mass is generally sufficient to complete the final hydration of the cement particles and achieve the optimum strength of the concrete.

Once the cement in the concrete has achieved its set, and the process of hardening is properly under way (generally something less than eight hours from the time water was first added to the mix), further wetting of the concrete will do nothing to retard the process of hardening, which has already been explained to be a chemical process and nothing whatever to do with drying-out. Nor can further wetting now disarrange the particles and spoil the density achieved for the concrete mass. Accordingly, in conditions of extreme exposure to sun or wind, it may be desirable to keep the concrete saturated by spraying with additional water even after the four-day period.

In Chapter 1 we mentioned *designed mixes* based on prior experience of the materials being used and the equipment installed in the mixing-yard from which the concrete was being delivered. Even without such experience it is nevertheless possible to specify within reasonable limits the proportions of materials that will consistently achieve a desired minimum cube strength: and whilst not going into this in any great detail, it is worth mentioning here the basic procedure so the reader will at least be aware.

First of course the materials need to be suitable; which for the cement normally means that it should comply with B.S.12; for the sand and stone means they should be of adequate strength, cleanliness, chemical inertness, impermeability, surface texture, size, particle shape, and grading to comply with B.S.882; and for the water means it should be clean and free from harmful materials such as clay, loam, or acids. Any town water-supply will do, and so will most river-waters unless polluted.

The proportions are then determined by pursuing the following procedure. The strength of the concrete will depend almost entirely on the water/cement ratio, *average* values at 28 days with ordinary Portland cement varying from about 20 N/mm^2 for a water/cement ratio of 0·7 to about 50 N/mm^2 for a water/cement ratio of 0·4. Obviously for safety we must concern ourselves with *minimum* cube values, so for the purpose of designing our mix we multiply the required minimum value by a factor so as to arrive at a suitable *average* value; and the factor will vary somewhere between 1·5 and 2·5, depending on the degree of control that will be exercised over the batching of the concreting materials, and the supervision that will be provided to watch over the placing and compaction of the concrete on site. However we also have to ensure that the *slump* of the concrete will be suitable for our task of placing, depending on whether the members to be concreted are small and much congested with reinforcements (requiring a high slump of up to about 150 mm maximum), or whether the sections are larger and with less reinforcement in which case a slump of about 50 mm will very likely suffice. Finally we have to consider whether the aggregate particles are rounded or angular and how they are graded, partly because the angular aggregates have a higher void percentage than rounded stones, and partly also because the angular aggregates do not settle into a solid mass as readily as the rounded stones for a given amount of mechanical compaction and lubrication (cement and water). Design tables are available which take into account all these features of water/cement ratio, slump, aggregate shape and grading, from which for any purpose it is possible to specify the proportions required of cement, sand, stone, and water. For normal work, the cement content is never taken less than 290 kg per cubic metre of finished concrete, except only where meticulous control is exercised over the water/cement ratio, in which case a value of 260 kg per cubic metre may be acceptable.

Whether the proportions of our mixture are determined by a method as described just above, or whether we are using *designed mixes*, or a *prescribed mix*, there will be no success in reinforced concrete work if the individual materials are not checked diligently at all times for suitability, and if the wet concrete is not placed in properly constructed rigid formwork, and adequately compacted to achieve a solid dense mass. Today the compaction of the concrete is normally done using vibrating tools; this enables us to manage with less water in the mix, and as the strength depends upon the water/cement ratio, this also gives us the chance to reduce the cement content as well, subject only to the limitations given in the previous paragraph. Mechanical vibration also helps speed up the work of spreading and placing, even with the employment of less labour. Thus mechanical vibration, which gives a better job, is normally more than paid for by the savings it achieves in cost of cement and manpower.

Up to this point we have been considering the concrete mainly from the point of view of its compressive strength, which is related of course also to its very much lesser tensile strength. Now we need to consider two further characteristics of concrete, namely *impermeability* and *bond.* Both relate to the successful performance of the steel bars within the concrete.

Impermeability in terms of concrete is relative. We have already pointed out why the best concrete we are likely to be able to achieve will still contain about 5 per cent of voids. For this reason, even the best constructed concrete water-retaining structures actually have water seeping through them at all times at a very low rate, but this evaporates from the outside face as quickly as it arrives, so that with good design it is never noticeable. It is true that after concrete has been submerged in water for a period of time, the water taken into the concrete causes the individual cement particles to swell so they fill the minute pores in the concrete with increased effectiveness, but it is the very absorption and replenishment of such water that represents the low seepage rate just referred to.

Concrete cover

What then is to happen to our steel reinforcing bars, which we have seen earlier in the book are always most effective when placed as near as possible to the faces of the concrete members? Invariably these bars are commercial steel with no protective surface treatment, and if left exposed to the damp would rust and deteriorate; indeed, in rusting, they oxidize to produce a far greater volume of material, and this pushes off the covering concrete so the bars would then become deprived of any concrete protection at all. Further rusting could then continue, until eventually the whole member itself might actually split: meanwhile all bond between the steel and concrete will have been lost. To avoid such hazards, it is of course important we take all the precautions described earlier to ensure our concrete is properly designed and compacted to achieve the optimum density; and in addition to this we take great care to keep the reinforcements in from the face of the member so they get a cover of concrete of 20 mm or

40 mm, depending on the degree of exposure, and the diameters of the bars. It is never wise to seek economy by adopting covers less than 40 mm except for members indoors or otherwise permanently protected against all effects of weather. With covers less than 40 mm it becomes increasingly difficult to place and consolidate the very concrete on which we are so utterly reliant for protection and permanence. Indeed, from what has been said before about impermeability, one might ask how absolute protection of the steel can be hoped for with as little as 40 mm cover. The answer is that with good concrete, even if some water-vapour does get through to the steel, this will be in such circumstances that air cannot be present with it too: and rusting occurs only as a result of free oxygen aided by the presence of the water, not in fact as a result of the action of the water alone.

If on the other hand a great excess of cover is provided, then the reinforcements will have to act in their member on a less effective lever arm than was assumed in the calculations for resisting the bending moments. This of course would lead to higher stresses in the steel, with greater strains and stretching, and therefore greater crack widths in the concrete. Thus it is important also to avoid the provision of too great cover; so we see the need for particular care and accuracy in placing the reinforcements to achieve the 40 mm cover stated above within a fine tolerance.

Bond

Let us go back now to the important matter of *bond*. In Chapter 3, when we were unfolding the basic principle behind all reinforced-concrete design, we stated the simple truth that in good design adequate safeguards are taken to ensure that the steel reinforcing bars are not allowed to slip within the concrete. The whole development of reinforced concrete as a material is dependent upon this premise for reasons that must now be perfectly obvious. If the bars in the bottom of our first reinforced-concrete beam were loose and could slip, the beam would fail in bending exactly as though it were unreinforced, since clearly the bars would not be contributing to any of the necessary tensile strength we have been assuming and basing our calculations on. Similarly, in our considerations of shear, the arching action in Fig. 6.4 could not take place if the longitudinal steel were not excellently secured within the concrete; nor would the links in Fig. 6.6 do their job if they were free to slip; nor would the bent-up bars in Fig. 6.10 be any use if, in carrying their force, they slipped freely in the concrete on arrival at the column. Similarly again in our chapter on columns, all the formulae we have given would be senseless if the concrete, in carrying the loads, did not transfer an appropriate share of these loads to the reinforcements, since otherwise the reinforcements would be idle as there is no other way that they could take up load and contribute to the performance of the members. All these examples show how we have been relying throughout this book on the bars not slipping in the concrete; and this absence of slipping is achieved by the capability of the concrete to develop a tenacious bond with the steel. We can consider this in very simple terms as follows.

If a straight steel bar is embedded in a block of concrete with the end of the bar left projecting at right angles from one face of the block, as in Fig. 11.2, it will be found that there is great resistance to pulling this bar out, depending upon a number of factors. The ultimate anchorage bond stress between the concrete and the bar is measured by the force P required to pull the bar out, divided by the area of the bar in contact within the concrete. If the length of embedment of the bar is l and its diameter is d, the total area of bond will be $A = \pi dl$ (mm^2), and the ultimate anchorage bond stress will therefore be

$$a = \frac{P}{\pi dl} \ (\text{N/mm}^2). \tag{11.1}$$

Now the ultimate tensile force in a bar of our 425 N/mm² steel is

$$P = \frac{\pi d^2}{4} \times (0\cdot 87 \times 425) \ (\text{N}).$$

So from equation (11.1) we have

$$a\pi dl = \frac{\pi d^2}{4} \times (0.87 \times 425) \ (\text{N}),$$

which gives

$$l = \frac{\pi d^2 \times (0\cdot 87 \times 425)}{4a\pi d}$$

$$= \frac{92\cdot 5 \, d}{a} \ (\text{mm}). \tag{11.2}$$

Different values for ultimate anchorage bond stress apply, depending on the grade of the concrete and the nature of surface of the bar, whether plain or deformed. For a 425 deformed steel bar in Grade 25 concrete a

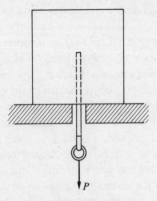

FIG. 11.2. Anchorage bond.

suitable bond value can be taken as $1 \cdot 9$ N/mm^2; therefore the bar will just be able to hold in the concrete up to the point of tension failure of the bar if we have an embedment length

$$= \frac{92 \cdot 5 \, d}{1 \cdot 9} = 50 \text{ diameters approximately.}$$

This is an important result, and known as the *anchorage bond length* for 425 deformed bars in 25 Grade concrete. A 25 mm bar would thus require an anchorage length of 50×25 mm = $1 \cdot 25$ m. If the concrete is not up to the 25 Grade quality, a lower value has to be taken for the bond stress, and consequently a greater anchorage length is required.

From what has been said it also follows that if two rods are lapped a distance equal to the bond length in concrete with plenty of concrete covering round, they will be as efficiently joined as if they had been a continuous bar or two bars welded together, and this fact is a great convenience in reinforced-concrete construction and repeatedly taken advantage of.

Experiments show that the anchorage bond actually depends on each of

(a)   the proportion of cement in the mix;
(b)   the strength of the concrete;
(c)   the surface nature of the bars;
(d)   the cover of concrete round the bars;
(e)   whether there are links or other circumferential reinforcements preventing the cover from splitting away.

The bond between concrete and steel is made up of two parts. The first part consists of a glue-like *adhesion*. Thus, if a shovel which has been used for concreting is left covered with concrete and put away in this condition, great difficulty will be found in cleaning the concrete off the shovel after a week or two. There is, however, another part of the bond resistance which is due to the *grip* of the concrete round the bar. The concrete shrinks on setting, and in shrinking squeezes the surface of the rod with a force which, multiplied by the coefficient of friction, represents the amount of bond due to this cause. The radial compression stress has to be resisted by circumferential tensions in the concrete surrounding the bar which are, of course, limited in stress by the tensile strength of the concrete and by the thickness of the concrete cover. This gives the clue to the reason why the adhesion depends on the strength of the concrete as well as on the thickness of cover, and the existence of links.

We have been talking here throughout about average bond stresses spread equally along the length of the bar, whereas in fact bond stress is not constant over the length of embedment, and in the case of our bar being pulled out would be a maximum value at the surface where the tension in the bar is its maximum. When a rod begins to slip at a section, the glue-like adhesion resistance breaks down and is lost while the residual bond due to the gripping may continue until the concrete cracks. Bond stress is greater for rods which are pushed than for rods which are pulled, because in the former case the effect of the push is to make the rod swell, while in the latter case the effect of the pull is to slightly reduce its diameter.

Cement types

For most normal reinforced concrete work, the cement used is *Ordinary Portland cement*. There is however a modified version of this, where, in the process of manufacture, the particles of the finished cement are ground down a little finer, with the result that the hardening process of the concrete becomes accelerated. This cement is known as *Rapid Hardening Portland cement*, and costs more than the Ordinary Portland variety to cover the cost of the additional grinding. The advantages of the use of Rapid Hardening cement are that the shuttering or formwork can then be released somewhat earlier, making for savings in the amount of formwork, time and cost: also, in hardening, extra heat is generated and this helps keep the hardening process going when working in winter conditions. The attendant risk that arises with the use of this cement is that the extra heat it produces increases the likelihood of minor cracking in the work; and for this reason it is probably wise not to use Rapid Hardening cement unless there is a real benefit to be gained.

In foundation work, one sometimes comes across soils, particularly clays, which contain sulphates in sufficient concentrations to attack concrete made with Ordinary or Rapid Hardening Portland cements. In such circumstances it is necessary to use *Sulphate Resisting Portland cement* instead. No different problems arise in the design or mixing or placing of concrete using Sulphate Resisting Portland cement. Ordinary Portland cement produces satisfactory concrete where the sulphate concentrations are not greater than 0·5 per cent in the soil, or 120 parts in 100 000 parts of the ground water, provided not less than 330 kg of cement are used per cubic metre of concrete, and provided also the water/cement ratio is not more than 0·50. Sulphate Resisting Portland cement concrete is satisfactory for sulphate concentrations up to 2 per cent in the soil, or 500 parts in 100 000 parts of ground water, provided not less than 370 kg of cement are used per cubic metre of concrete, and provided also the water/cement ratio is not more than 0·45. Whichever cements are used in the presence of sulphates, the results will only be satisfactory if the water/cement ratio of the concrete is closely controlled and the concrete is properly compacted all as discussed earlier. It is the sulphate solutions in the ground water which destroy the cement as they percolate through the concrete; and it is only by having the concrete so dense that the ground solutions cannot work their way through the concrete with sufficient ease to renew their sulphate concentrations that successful permanence can be achieved in such circumstances.

Concreting in cold weather

In Chapter 1 we mentioned how the strength of concrete increases with age, particularly over the first 28 days of its life. This rate of increase of strength is very much influenced by temperature. Concrete at $2°C$ hardens at only about two-thirds the rate of concrete at $15°C$, and when the temperature falls to freezing point the hardening process stops altogether, unless there is sufficient heat being generated by the chemical reaction to keep the process going. Nevertheless hardening will re-start once the temperature rises above zero again, though damage to the concrete from the expansion of ice may already have been done.

For these reasons, it is best that concrete should not be placed when the air temperature is below $2°C$; and for practical speedy work, enabling the shuttering to be removed after a reasonable interval, it is generally advisable to heat the concrete materials to a temperature of 15°C whenever the air temperature is below 5°C.

The process of heat generation from the chemical reaction of the concrete hardening has its advantages in certain circumstances; in members that are 500 mm thick or more, this heat may be sufficient to assist in the protection and rate of hardening of the concrete, particularly if the concrete is insulated by covering with straw or sacking or other suitable material to hold the heat in. However with thin concrete members, the heat from the chemical reaction is generally insufficient to build up against the temperature losses from the proportionately larger exposed surfaces; and these are the circumstances in which it is specially important to heat the materials and protect the finished work from frost attack.

It is hoped these general points about the behaviour and problems of concrete give some insight to the fickle variabilities that crop up in practice. Such a short chapter could hardly seek to do much more. But having tried earlier to express the fundamentals of design in easy terms, it seemed important to remove any impression that reinforced-concrete construction was a simple foolproof matter; because certainly it is not.

Perhaps good construction can be compared to golf where the objectives are obvious, and the process, to the onlooker, seems so straightforward. Yet the more one gets involved in either, the clearer it becomes the whole thing is made up of many tiresome little parts, the effects of which together are out of all proportion to their apparent significance taken separately. Watching at a construction site, the entire performance seems casual and unfussed; the concrete goes in grey and easily, a display of elegant simplicity. Let's just hope there were a few careful practice swings behind the club-house first, so the first drive isn't sliced into the spinney over by the 14th green.

Index